Whale Falls

An exploration of belief and its consequences

by Cecil Bothwell

Brave Ulysses Books
2010

Whale Falls
EAN-13: 9781450555036

Brave Ulysses Books
POB 1877
Asheville, North Carolina 28802
braveulysses.com

also by the author
•Gorillas in the Myth: A Duck Soup Reader
2000/second edition 2008
•The Icarus Glitch: Another Duck Soup Reader
2001
•Finding Your Way in Asheville
2005/2013
**•The Prince of War: Billy Graham's Crusade for
a Wholly Christian Empire**
2007
**•Garden My Heart: Organic Strategies for
Backyard Sustainability**
2008
**•Pure Bunkum: Reporting on the life and crimes
of Buncombe County Sheriff Bobby Lee Medford**
2008
**•Can we have archaic and idiot? A collection of
fictitious tropes**
2009

•She Walks On Water: A novel
2013

Whale Falls

for Susan
against all odds
and in spite of myself

Contents

"Some say the world will end in fire,
Some say in ice.
From what I've tasted of desire
I hold with those who favor fire.
But if it had to perish twice,
I think I know enough of hate
To say that for destruction ice
Is also great
And would suffice."

— robert frost

"I wish I had a river so long
I would teach my feet to fly
Oh I wish I had a river
I could skate away on
I made my baby cry"

—joni mitchell

windfall

The clan gathered on the beach.

Children were silent and their parents murmured quietly while all together they stared at the shimmering behemoth which lay stranded by the outflowing tide.

Whales were not unknown to the people of the coastal forest. They had occasionally spotted pods swimming near the coastline, had noted the spout or spume when the great beasts breathed, and had witnessed a sudden breach or the slap of massive tails and fins. But the Neanderthal were not seafaring folk and, in fact, had not yet learned to include fish in their diet. The sight of such an enormous beast stretched out on the strand, feebly twitching and breathing out in labored gasps presented an entirely new tableau.

An experienced hunter advanced and touched the whale's side with his wooden spear. There was no reaction. He pressed harder and a coconut-sized eye eased open to fix him for a long moment in a clearly sentient gaze before falling shut once more.

When the clan returned the next morning the creature no longer breathed and chunks of its flesh trailed in the surf, torn out by sand sharks which had dug into the carcass during the intervening high tide. Gulls were picking at shredded wounds and crabs scurried and nibbled at the margins.

Two hunters drew stone knives and carved into the whale body, tasted cautiously, then took full bites of the succulent red flesh. As the unfamiliar meat was passed from hand-to-hand and savored mouth by mouth others joined in the butchery. Later still, chunks of whale meat, some bearing sizable slabs of blubber, were impaled on sticks and roasted over a cook fire where the dripping fat set off leaping flares as it sizzled into the flames.

The Neanderthal were familiar with the various properties of animal grease and quickly understood the fuel value of the beast delivered to their shore. They gorged on whale meat for days thereafter, until the flesh began to putrefy, and used the blubber for much longer to stoke cook fires and torches and to work into animal hide to render it waterproof.

One among their number carved a whale tooth into the rough shape of the beast and placed it near the mouth of the communal cave, a remembrance of the clan's great good fortune or its blessing by the gods.

Perhaps the first of our predecessors to utilize whale products were not Neanderthal but Cro-Magnon or *Homo habilis* or *Homo erectus* or *Paranthropus,* but at some time in the ancient past we began a cultural, ethical and utilitarian dance with our cetacean kin.

Mesolithic sites excavated in northern Scotland have yielded whale bone artifacts dating to eight thousand years ago. The earliest evidence of whaling as an intentional activity has been traced to Korea, between 6000 and 5000 BCE, and in Norway, about a thousand years later. Bronze age middens and mounds in many coastal areas around the globe indicate widespread use of whale bone by about 3000 BCE.

Like other mammals, whales and dolphins are descended from walking fish who crept out of the sea to populate the land masses of our planet. But we remained ashore while cetaceans returned to the water, there to evolve to include the earth's largest creatures. Along the way many species developed thick layers of insulating blubber to shield them from the cold and pressure of life in the frigid deeps. Their enormous size and bountiful fat made discovery of beached whales an inestimably rich, if unpredictable and rare, resource for early humans.

This is a story about our relationship with these biggest-brained creatures on earth, our concept of consciousness and mind, our deepest superstitions and best science, the results of our widespread dependence on whale oil followed by fossil fuels, and of possible repercussions of blind faith and unquestioned technology on our collective future.

1. feeling

Some whales are utterly enormous.

One can't help but wonder what a one hundred and ninety ton creature feels. What feedback do they receive from the distal portions of their vast epidermal sheath? They are inured to cold, surely, or at minimum less sensitive to extreme cold than you and me. We know they tend to be well insulated with blubber, of course, but that is interior to the skin. While it may serve to keep their innards warm in the frigid, heat sucking sea, there is still that outer layer of whaleness that interfaces with the aquatic realm from the tropics to the poles. Skin is the sensory organ that collects data about the air-ocean puddle in which we earth-folk swim or walk or fly.

In our own skins we have nerves that react to temperature, to the tiny needley tube inserted by a mosquito, to a lover's touch, to a bucket of ice water in the night.

How sensitive is whale hide by comparison? Is the rush of water across her skin as pleasurable as my warm shower or your sitting beneath a cascade in a mountain stream? Is the penetration of a harpoon into a cetacean more akin to a mosquito bite on one's arm, or a spear in the gut? Fat doesn't contain a wealth of nerves, so if the skin is relatively numb, how far in does the barbed weapon have to go before she feels pain?

Or is the pain of violence really more about violation of one's envelope than it is about sensation? It's hard enough to claim certainty about how or what other human beings feel. The basis for interspecies empathy is largely conjectural.

Not that most humans have thought long or hard about whales' pain. Before the last years of the last century, most who considered whales at all were chiefly interested in converting them to cash. At one point whaling was the fifth-largest industry in America. Collateral damage has never much mattered where it stands between men and wealth—whether it's cancer and black lung in coal mining communities, children blasted by cluster bombs and land mines laid in oil fields, or gang victims in the misguided and fraudulent war on drugs. The pain of whales, poor or non-white children, and those defined as criminals rarely affects public policy, no matter how they might climb our personal heart charts.

When the discovery that whale blubber could be rendered to useful fuel was combined with improved navigation and the construction of large ships, we started on the road to Exxon-Mobil, the wars in Iraq and Afghanistan and global climate change.

"Thar she blows," indeed.

It all started with beached whales.

For reasons largely conjectural, whales beach themselves but there's strong anecdotal evidence that human technologies trigger some beaching. For example, the U.S. Navy's new high powered sonar is a particularly potent source of intense sound waves. Cetaceans are a sonically attuned, auditorily nuanced family. Think classical music aficionados plunked into a heavy metal concert, times ten. Who knows, really? Times a thousand? Times ten thousand?

Leave the concert?

Leave the ocean?

But "natural" whale beaching far predates intrusive human technologies. Recall Jonah, whose myth could easily have been swallowed by humans familiar with great "fish" on their beaches. There's some evidence that whales in poor health run themselves aground and perhaps that's not much different than the practice of Inuit elders who go out on the ice when their time arrives.

In any event, as described in the opening vignette, the arrival of huge mammals on the shore was a great windfall for our ancestors who ate the flesh, used the bone for tools and rendered the fat. Whale oil became an important but sporadic source of energy. Various peoples hunted the giants starting several thousand years ago, but their use of whale products on a subsistence level didn't contribute to the evolution of our industrial and petroleum economy.

Subsistence hunting remained the rule until larger boats and dense population centers combined to create the means and the demand for large scale harvest. English, Dutch and Norwegian whalers began to work the waters of the north Atlantic in about 1600. Then, in the 1630s, the Dutch became the first Europeans to successfully conduct commercial whaling in U.S. waters.

The financial reward for successful whaling was enormous. A single right whale carcass towed to shore in Rhode Island in 1662 reportedly garnered more cash than a whole farm could earn in a year. In addition to the oil, right whales were "right" because their bodies floated when they died, making them easier to haul, and because their feeding apparatus consists of baleen, formed in a mesh with which the mammal strains sea water to sift out morsels of food.

Baleen becomes malleable when heated and then keeps whatever shape it is cooled into, and became the equivalent of today's plastic in early manufacturing. Its primary use was for corsets, but it found use in numerous other applications, from combs to shoehorns to umbrella ribs and fishing rods.

In a 17th century version of "The Graduate," we'd hear the following exchange:

Capt. McGuire: I want to say one word to you. Just one word.

Deckhand Benjamin: Yes, sir.

Capt. McGuire: Are you listening?

Deckhand Benjamin: Yes, I am.

Capt. McGuire: Baleen

Deckhand Benjamin: Just how do you mean that, sir?

Capt. McGuire: There's a great future in baleen. Think about it. Will you think about it?

Deckhand Benjamin: Yes sir.

Nantucket soon became the epicenter of global whaling due to its geographic location—and offshore whaling was dominated by Nantucket until the early nineteenth century.

As related by Caleb Crane in a book review in the *New Yorker* (July 23, 2007):

> *This involved the island in a certain amount of un-American activity. War interfered with profits, so, when conflict with Britain loomed, Nantucket tried to stay out of it. In 1775, when Britain moved to restrict New England's trade and fishing rights, Nantucket won a special exemption by pleading pacifism and loyalty to the Crown. The neighbors on the mainland took note, and, once the Revolutionary War broke out, rebels seized flour and whaleboats from Nantucket and put the islanders under an embargo, suspecting them of trading with the enemy British. Some Nantucketers did indeed intend to trade with the British, and a few went so far as to base their whale ships in the Falkland Islands. Others, who stayed, won permission from Massachusetts authorities, in 1779, to ask*

*British military officials not to raid them, a bit of
diplomacy that, as an American general pointed
out, was traitorous.*

By this time, the street lights of London and Paris
were fueled with whale oil, creating an enormous demand
for the stuff. The British, anxious to build their own
whaling fleet, consequently imposed high tariffs when
trade with America resumed after the war although it was
far from clear that their own production was sufficient. In
response, as noted above, some Nantucketers declared
neutrality (essentially seceding from the U.S.), others
moved operations to Nova Scotia or the Falkland Islands,
and a few settled in Dunkirk, welcomed by the French
with their own market to appease.

During the War of 1812, Nantucket formally
declared neutrality in order to maintain trade with the
nation's enemy.

Oil money, then as now, superseded patriotism for
many of those with a big stake in the trade. Halliburton's
sub rosa trade with Iraq during Dick Cheney's tenure as
CEO is an example. As reported in the *Washington Post*
(June 23, 2001), despite Cheney's denial, "According to
oil industry executives and confidential United Nations
records, however, Halliburton held stakes in two firms
that signed contracts to sell more than $73 million in oil
production equipment and spare parts to Iraq while
Cheney was chairman and chief executive officer of the
Dallas-based company." This represents an explicit
violation of the trade embargo imposed on Iraq following
the first Gulf War.

When   Halliburton   moved   its   corporate
headquarters to Abu Dhabi in 2008 the corporation was
simply, like Nantucketers of old, declaring its true
allegiance.

Traitorousness.

There's a great future in traitorousness.

fire

Owls conversed after nightfall

The stars were bright and there was just enough breeze to kindle a soft sussing of trees. We were camped at the Dennis Cove Campground in the Cherokee National Forest outside of Elizabethton, Tennessee, on a Thursday evening. The recreation area was deserted. We parked and let the cats out for an evening stroll and built a fire in the pit provided. Barred owls set to with their "Who cooks for you? Who cooks for you? Who? Who?"

Taken all together it was a perfect camp evening.

At about noon the next day people began to arrive. Party people. Their radios were blaring country music and loud voices swelled to fill the area. At about 4 p.m. a real rush of arrivals commenced until all the fifteen or so campsites were packed with cars, campfires smoking, barbecues sizzling and boisterous partiers crowded around picnic tables and coolers. Horse shoes clanged against stakes and each other in the pit. Kids raced around on bikes. The odd thing was that there were no camper units and few tents.

A very loud, very fast automobile with street slicks on the rear, chrome wheels and "Simonize that hurt the eyes" came roaring into the site and a twenty-something man jumped out and opened the trunk. A line formed and we could see people walking away carrying quart canning jars of clear liquid. The trunk was closed and those remaining in line turned away while the young man assured them loudly, "I'll be back." Then off he roared.

Soon the party got louder and in about an hour the delivery car returned. Meanwhile, I sat at our table and played my guitar while I watched our neighbors get steadily drunker as they passed around jars of white lightning. Some simply passed out, leaning against trees. Others roared on into the night and the celebration didn't seem to quiet at all, long into the wee hours. Some few were still going strong when we awakened the next morning.

All day Saturday the party continued. A couple of fights broke out, but the combatants were so sloshed that their punches weren't connecting, and they generated far more laughter than angry shouting. More and more of the participants could be seen napping, stretched out on picnic tables or tipped back against trees as the afternoon warmed. By nightfall most of them were back on their game though, the delivery car made more runs, and an older fellow who introduced himself as Jeff came over to ask me if I'd care to come over to their site and play a few songs.

Soon after my arrival I was offered a jar and helped myself to a couple of sips over the course of the evening. The sing-along was spirited. The likker was potent. I slept early and then slept late Sunday morning, waking up with a terrible headache. When I'd made coffee and found my bearings, I noticed that Jeff was the only survivor up and about at the next site, so I went over to chat and allowed that it had been pretty powerful and pretty raw liquor to a newcomer like myself.

"You know, if you'd age it in barrels, like the bonded stuff, it would be just as smooth as Black Jack Daniels," he told me. "Of course, none of 'em want to hold onto it that long.

"My Daddy used to brew up a batch now and then,

and the slow part was making the beer, waiting for it to ferment. So Daddy, he'd toss a shovel-full of litter out of the chicken coop in with the mash to speed it up."

Incredulous I asked, "And he drank it?"

"Hell no, son. He *sold* it."

Jeff went on to tell me that his Daddy was clever enough to never get caught. "The main thing, when you're runnin' it through the still, is to do it when the wind is right. Folks smell smoke up a holler where there's no house, and they know somethin's up. 'Course the neighbors knew anyway. They were all customers."

The inclination to alter consciousness through ingestion of alcohol, caffeine, nicotine, cannibinoids, opiates, ergolines, tryptamines and alkaloids is apparently older than civilization. Discovery of late Stone Age beer jugs has established the fact that intentionally fermented beverages existed at least as early as 10,000 BCE, and some authorities have suggested that beer preceded bread as a staple.

Indeed, there is strong evidence that drug use is prehuman. Reports of caffeine-dependent goats and nectar-addicted ants, accounts of mushroom-loving reindeer, intoxicated birds, and drunken elephants abound. Only humans, possessed of self-awareness and written language, have created proscriptions on the drug habits of others. To that end we have locked up or killed bootleggers and rum runners, international kingpins and street dealers, and waged wars from the purely metaphorical to the onslaught of armed armadas.

2. money

Wild Arctic char have delicate flesh.

I had been fishing somewhere east of Fairbanks, Alaska, and I returned to camp with three Arctic char and no wallet. I discovered the absence while cooking the pale pink fillets.

The wallet contained cash—about $350—my entire travel fund for the two months remaining in our itinerary, as well as my drivers' license, my social security card and my fishing permit. The July evening was fading toward the dusky grey that passes for night during Alaskan summers when I noticed my empty hip pocket, so I waited until morning to retrace my steps. Unlike most waterways near population centers, wilderness rivers rarely have well-worn footpaths along their banks and I had done a fair amount of bushwhacking during a long afternoon of angling along a mile or more of a crystalline waterway.

No luck.

Susan got right to the point. "I have enough money to keep going. What are you going to do?"

We had only been together for five years at that time and her disinterest in functioning as meaningful partners still came as a shock. That this distancing involved considerable denial on my part didn't really sink in until shortly before she died, and even today I am repeatedly surprised by my own journal entries from those early years. The difficulties I finally acknowledged two decades later were there from the start, and I took note of it. Literally.

It seems we only hear what we want to hear, even when we are talking to ourselves.

"So you're going to leave me here?"

"How are you going to pay for gas? Maybe you can hitchhike."

It immediately sunk in that we were traveling in *her* Volkswagen (which *we* bought) and that she would feel entirely within *her* rights to ditch me there. I was completely at *her* mercy. The matter of ownership and title and personal trust would eventually deliver one of the bitterest lessons in my life. But then, as now, my attachment to material goods was minimal and the fact that she held title to *our* car had seemed irrelevant. My mistake.

"Will you take me to Homer?" I asked.

We'd heard from others during our travel up the Alaska highway about the canneries in Homer where college kids found summer jobs where they reputedly made good money.

"You can leave me there," I added.

"You'll have to pay me back your share of gas and food."

"No problem."

So we continued our journey, on *her* nickel as she kept reminding me, and stopped off for a few days in Denali National Park en route to the Kenai Peninsula south of Anchorage.

Hiking there I saw my first grizzly. She was galloping across a slope on the opposite side of the valley, chasing a ground squirrel. I had no idea bears could move that fast, and she was magnificent. (Two cubs trailing the hunt were proof of her gender.)

As the, what? eight hundred pound?, bear galloped across the rock strewn slope sunlight flashed from her golden fur. She whirled on a dime, and again, and again, and then began digging furiously where the squirrel had gone to ground. Great clods of dirt and gravel flew in all directions while the cubs rolled and tumbled in mock battle nearby.

Much later, down south of Anchorage we'd watch a half dozen grizzlies scooping salmon out of a frigid river. Some lunged in and grabbed the squirming fish in their teeth, others batted the salmon up onto the bank, then pounced on the flopping prey. Coming up out of the water the massive beasts would shake their entire selves sending off buckets of spray. By all appearances the cold didn't faze them at all. They appeared to be having great fun—though, of course, it was all about survival, building up fat stores for the coming winter.

Current reflection on those bear sightings takes me back to my earliest travel venture with Susan, in 1976, when we left Florida to seek some other fortune. We were traveling in *her* van atop which sat *her* canoe, two items she had kept when she separated from her husband. (She also refused to pay for a divorce and died married to him, a fact that I conveniently managed to forget until I was sitting in a lawyer's office twenty-seven years later.) At that point, although we were more or less romantically involved, she had made it clear that I did not necessarily figure into her future plans and she was merely giving me a ride north.

I was leaving behind a dissolved marriage and hoping for a new start in music or writing, somewhere north of Florida—whose endless summer had stultified in the thirteen years of my residence. Seasons beckoned. My hopes were high and my typewriter and guitars and I had a ride north. That was enough for me.

One night, camped in the Great Smoky Mountain National Park, Susan regaled me with bear stories. I'm not sure where she picked them up, but she claimed to know a great deal about human attacks by black bears and offered up lurid details of people mauled and massacred in their sleeping bags and of children dragged off and consumed.

I'd been an outdoors sort since childhood and spent a great deal of time camping in Scouts and with my family and friends, but bears weren't a particularly familiar subject. Even as a camp counselor teaching Nature, Insect Life, Reptile Study, Mammal Study, Bird Study, Botany, and Soil and Water Conservation merit badge classes, I'd never seen a bear in the wild and had heretofore regarded black bears as benign. But I could believe I might be wrong. When it was time to turn in, I left the van to take a pee in the woods and returned to find the doors locked.

I knocked. No answer.
I knocked again. "Susan, let me in."
I raised my voice, "Susan?"
"Sleep on the picnic table," she replied.
So I stretched out on the picnic table while the bears that populated her stories cracked branches and snuffled in the woods around about. I wasn't exactly frightened, but neither could I sleep. A couple of hours later she opened the van and called me in, while evincing great delight at her practical joke.

I would never really learn that practical jokes were one of her special joys, even when I was occasionally enlisted and not infrequently the butt, and the pain sometimes cut to the core.

Humans are cooperative animals, it seems, and going along to get along is usually the order of the day. It makes community possible, and permits us to live with emotional and physical abuse, or even slavery.

ice

Eagle Scouts face final judgment.

At least I did. I'd guess the process hasn't changed a great deal since 1965, although I've heard that they've added a requirement for a community service project (and more power to that!). You complete the requirements—chiefly acquisition of twenty-one merit badges—and you go before an Eagle Board of Review. Three adults sat in judgment of your worthiness and grilled you; partly to confirm that you had completed the work, and partly to impress you with the seriousness of the honor you were about to receive.

There I was. Age 14. Requirements more than fulfilled (I had 24 merit badges.) And one of my interlocutors was suddenly all panties in a twist about God.

"I see you haven't earned the God and Country Award." This was a medal earned through church service and other requirements. Researching it today I see that it has been replaced with myriad other religious awards and badges,

"No sir."

"I like to see a Scout earn the God and Country Award before he earns the Eagle rank."

"Yes sir."

"I would like to know that you plan to earn the God and Country Award soon."

I'm thinking, "Oh, hell. This prig is going to deny me my Eagle badge if I don't agree. My achievement of the award is hugely expected by my father, an Eagle himself. I have really worked hard and this jerk is making an entirely extraneous demand, with no grounding in the official rules. The injustice is almost too huge to bear. But he's an adult. I'm a kid. I don't talk back to adults and I'm very respectful of authority. There must be some way outta here, said the joker to the thief."

"I'll try," I replied.

"Do you attend church?" my inquisitor demanded.

"Sometimes," I sort of lied. After all, my Boy Scout troop meetings were held in a church.

No, I mean, I lied. Because I knew what he meant was not what I meant, and I hadn't been to a church service in years, outside of Scout activities.

"Well, I'd like to know that you attend church regularly and that you will work on earning the God and Country Award."

"I'll work on that, sir."

And so I was awarded my Eagle Scout badge, an award that is touted as signifying good moral character, by dint of a lie. I probably could have fought back instead, demanding that I not be held to account for a non-existent requirement. Maybe if I'd told the truth I still would have gotten my badge. I console myself that my falsehood emerged from my sense of powerlessness as a child facing an adult. But I knew damn well I had violated my Scout oath.

My main take-away message was that pretense was more important than performance in parts of the adult world. In the aftermath it became more and more clear to me that I didn't much believe in gods or God and that I had been coerced to lie, and perhaps worse, that Scouting's highest honor was very cheaply purchased. In the end the rules didn't matter.

My personal ethics were still being formed at that point, and as I puzzled through my notions of right and wrong I realized that my personal ethic was worlds away from that of the Boy Scouts or its putatively religious leaders. Lies were lies. I wouldn't permit "grown-ups" to twist my ethics again. I wouldn't let myself lie.

On the other hand, I wasn't one hundred percent certain about religion and God. Toward the end of my senior year of high school, three years later, I fell in with a bad crowd who tempted me through the following summer. Campus Crusade for Christ had set up shop in my town, with a very charismatic young man who managed to make religion seem cool to several of my friends. I joined in and attended weekly meetings, bought a copy of a modern translation of the New Testament and was duly saved. A couple of months into college the following fall restored my rationality and I recognized the experience for what it had been—typical teenage seeking for acceptance within a peer group that seemed "with it."

I decided I could be cool on my own.

3. crabs

Fishing was off in Homer that year.

While the stories of seasonal workers making good money by laboring long hours processing the commercial catch were true enough, that all comes down to fishermen's luck. In that month of that year the boats returning to port were running high for lack of take.

I managed to find a job in a crab packing house as a butcher. Decked out in rubber boots, rubber gloves and rain gear I worked eight hour shifts under a constant spray of water. A fork-lift operator shoved pallet-sized boxes—filled three feet deep with milling, crawling, shifting king crabs—into the slaughter room. In front of me stood a blade on a stanchion, essentially a dull broad-axe welded business-end up on a steel pole.

I grabbed an enormous crab by its legs on each side, brought it down hard across the blade and tossed the legs onto a conveyor while the carapace and guts fell to the floor. The legs were still twitching as they disappeared into the factory tunnel. And again. And again. And again.

A half hour into each day I'd be standing ankle deep in dead crab parts, my rain gear splattered with guts and gore that ran down under the constant spray. When the footing got treacherous I kicked the pile toward a hole in the floor whence emanated the constant grinding growl of the machinery which would render the innards and shells into crab meal. By the third day I was somewhat inured to my role in the killing business, death and profit mingling in my own and the packing house's favor. And after all, it wasn't like I didn't eat crab myself.

But the scale of the slaughter was boggling. I smashed hundreds and hundreds of living, struggling creatures across the wedge-shaped blade. And again. And again. And again. Nothing I experienced before or since has quite as viscerally impressed me with the scale of human consumption. All that death to feed all that life. And again. And again. And again.

Nor is the death limited to the crabs. Alaskan crab fishing is ranked as the most dangerous occupation in North America, and the fatality rate among fishermen is about ninety times the fatality rate of the average worker. The major causes for death are hypothermia and drowning and fishermen are frequently severely injured or killed by heavy equipment operating on icy decks while ships toss in roiling seas.

When I was there, in the summer of 1981, it was just a year past the peak of the king crab industry when Alaskan fisheries produced up to 200 million pounds of crab per year. By 1983, the total catch had dropped almost sixtyfold, so I was witness to the beginning of the crash.

Today, Alaskan king crab fishing is only permitted during the winter months in the waters off the coast of Alaska and the Aleutian Islands. The commercial harvest is limited to a very short season, about four days, and the catch is shipped worldwide.

Reflective of the crashing numbers, after my fourth day I was laid off. The packing plant was out of crabs. But I'd made enough money to pay my half of current expenses and we headed back north toward Anchorage with some hope of finding construction work.

As stated, I was no innocent concerning crabs. Five years earlier, right at the beginning of our life together, we had camped on a beach on Cedar Key, then a quiet fishing village on Florida's west coast. (Later to morph into an upscale resort destination favored by Jimmy Buffet and friends.) The bay was speckled with buoys marking the trap lines of local crabbers and I bought a crab trap at the local hardware to try my luck. I baited it with chicken necks and anchored it with a chunk of coquina rock. Twice a day I paddled Susan's canoe out to check for dinner. Nothing.

Susan was personable and fun-loving, always ready for an adventure and one afternoon she befriended a local crabber named Larry who offered to take us out into the Gulf of Mexico on his regular rounds. We met him on the dock at 5:30 the next morning and set off into the fog in an 18-foot open boat, headed for his trap line fifteen miles out to sea. The haze lifted as the sun rose and the coast first came into view, then gradually shrank to a thin line and finally vanished. Not long after we lost sight of land the six cylinder motor, which we soon learned had been salvaged from an old Ford Falcon, coughed and died.

Larry laughed, said it happened all the time, and tossed us each a beer after cracking one himself, and pulled up the engine cover to have a look, muttering something about a damned carburetor.

I popped the lid and tapped my can against Susan's. "Breakfast of champions," I toasted, and took a swig.

We drifted for half an hour or more before our captain solved his carburetion riddle and fired up the motor again. Another half hour on we were pulling his traps, well-laden with blue and stone crabs as well as unmarketable spider crabs, odd mollusks and stray fish, the traps all draped about with sea weed.

Blue crabs are sold whole, so those went straight into a holding tank, but stone crabs are only valued for their claw meat. In order to preserve the resource, fishing regulations require that crabbers only take one claw, leaving the other so the released creature can feed itself and grow a new arm, blessed as they are with the faculty of limb regeneration. In addition, the law required that claws be of a certain size, again to preserve the resource and create a level playing field for the harvesters.

Larry cheerfully snapped off both claws of each stone crab he caught, explaining that the authorities had no idea what they were talking about and that crabs were perfectly able to feed themselves with no claws at all. He assured us that all of his fellow crabbers did the same. though the plentiful number of one-armed crabs we retrieved suggested that at least some of his competitors were law-biding. Armless, all the stoners went back over the side.

The measuring device for determining whether claws were of legal size came in for regular use. Our guide tossed all of the legal claws into a large red cooler atop the engine cover while the undersized claws went into a separate cooler which he thoughtfully stashed under a tarp at the bow of the boat.

We re-baited each trap with chunks of mullet before lowering it into the water, which seemed surprisingly shallow for as far out to sea as we had travelled. No more than twenty feet deep, tops.

The sun was nearly overhead as we started back in to port and soon enough we were intercepted by a speeding patrol boat bearing the insignia of the Florida Fish and Game Department. The officers greeted Larry by name and asked him to hand over his cooler for inspection. They expeditiously measured the stone crab claws he had taken, told him he seemed to be having about the same luck as other crabbers they'd seen that day, and sped off.

Our guide opened a fifth beer and toasted the departing boat. "Just doing their job," he grinned. "I'm just doing mine," and we continued on our way. By the time Cedar Key hove into view he was well into his second six pack and I was feeling just a little surprised he'd found his way back home. Far less surprised, I note in retrospect, than I would be fifteen years later, on a return visit to the island, when I learned that Larry had died of alcohol poisoning before his 45th birthday. One bender too many, they said.

Back on the dock Larry gave us a bag full of illegal stone crab claws, urging us to eat them soon and not get caught, then inviting us to join him a local bar that evening to shoot some pool.

We had just finished our illicit dinner when a patrol car pulled up in our campsite and a sheriff's deputy climbed out. Susan was in the van, changing clothes and I stood and walked over to meet him.

"Good afternoon, officer. What's up?"

"Is that your canoe?" he asked, gesturing at the boat locked to a tree some fifty feet down the shore.

"Yes," I responded.

"Word is that you've been stealing crabs from the boys traps. And they're some upset. They sent me down here to arrest you."

"No sir. I only checked my own trap. That's it there next to the canoe. I bought it a few days ago at the hardware, and I didn't catch any at all."

He looked toward the campfire. "There's crab shells."

"We went out with a crabber named Larry, helped him with his lines today. He gave us some for dinner. He's a generous man.""

The deputy grinned, "Larry. So you've met Larry. Sorry to bother you, but I'm just doing my job. Maybe it'd be best if you moved along, though. The boys aren't catching many crabs just now and they don't want competition."

Just then Susan stepped out of the van. "What the hell are you hassling us about?" she called at the officer who stopped and turned. She strode up to him, got right in his face and demanded, "You have some kind of trouble with people camping?"

The cop stepped back and spoke to me over her shoulder, "I believe you need to call off your little lady."

"Susan," I said, "It's okay. Let it go."

"You think you have all the fucking authority in the world, don't you!" she shouted, "Well you can go to hell! We'll camp here as long as we damn please and you can't tell us any different!"

He spoke to me again, "Your little lady is going to wind up in jail if you don't call her off."

I stepped between them. "Susan. Let it go."

"You can go to hell too!" she spat at me as the deputy crossed to his car, shaking his head. "I am not your fucking little lady!"

"I didn't say that. Let it go."

"He goes on the list," she said, going back to the van and pulling out her purse. She produced a folded piece of paper and proceeded to write down a name and badge number (which I've long ago forgotten).

She never explained to me why she went off on that deputy, but I did learn some years later that the list was of people she intended to kill if she ever contracted a fatal illness.

Maybe a lot of people have those lists. Thirty years later I would learn that I was on the list of a North Carolina sheriff who told his cohort that he would take care of the reporters who had plagued him if he were ever informed he was terminal. Then I learned he'd undergone surgery for a cancerous kidney. Though he's in prison now, I suppose I still am on that roll.

The last time I saw Susan's death list there appeared to be more than two dozen names, though, when I thought about it at all back then, I tended to regard it as simply black humor.

We shot pool with Larry that night, danced with him to oldies on a still older juke box, drank way too much and stumbled back to camp well past midnight.

Susan had come up with a new destination, so we packed up and headed down the road after one more night on the Cedar Key beach. She wanted to be sure that locals understood we weren't moving on in response to the deputy, she explained to me, but because she had other plans.

4. breach

Ιt had become my habit to carry tools.

When I traveled I packed box-, socket- and torque-wrenches, screwdrivers, pliers and such to keep my old vehicles running; rope and come-along and crowbar and flatbar to unstick what got stuck; and my trowel, brick hammer and level so I could pick up masonry work as needed.

In Anchorage I searched for work laying brick or block or stone but soon learned that unit masonry is uncommon in a land where frequent earthquakes shake mortar joints loose. Construction ran to concrete and steel commercial buildings that could flex with the tremors or else wood-framed homes which are equally forgiving. More than once that summer I awoke in the night to a world swaying on jello stilts. Limited earthquakes are actually kind of fun.

Having dropped in on a few job sites and admitted to knowing a little about carpentry, I soon landed a job. "A little" was accurate enough, having used a skill saw and various hand tools now and again since the pre-teen days when I earned Home Repairs merit badge. But I knew very little in terms of real carpentry. I was told to arrive with my own hammer and that the rest of the necessary tools would be provided.

The only claw hammer available at the only hardware I could locate was a 32 ounce Vaughan, twice as heavy as a standard carpentry hammer. I swung it, feeling doubtful, but the salesman assured me I'd be glad of its weight. Thirty years and three replacement handles later, I still love that tool and I still get ribbed about its size on job sites. There is a whole lot of truth in the old adage, "Get a bigger hammer," and if you ever have the need to drive framing nails all day I'd advise you do the same. The strain of swinging back a two-pounder fades in comparison to the ease with which it sinks 16 and 20 penny nails into dimension lumber, like sharp toothpicks slipped into warm cheddar chunks.

A few days later my new boss asked me if I could read blueprints. With that answer I could be more forthright. Between an eighth-grade mechanical drawing class and several years of masonry work subcontracting foundations and buildings I was well versed in reading building plans.

"Well, you're my new foreman," he announced, explaining that he had another job to start and that none of his other carpenters could read prints. He assigned me to a house in Eagle River with two carpenters on my crew. "Do you know a helper?" he asked.

Susan and I had worked together building chimneys and fireplaces in New Hampshire, and I guessed she might want work, so I said, "Sure."

The boss said, "Send him around."

While he raised his eyebrows when a woman showed up on the job, he hired her and weeks later admitted that he had been hesitant but was well pleased. She was a hard worker and we were a good team—better at working together than living together, really, though I wouldn't come to terms with that relationship for many years.

I obtained an Alaska driver's license and a library card and checked out books on carpentry. Each night I would read about the tasks we had to do the next day so I wouldn't seem too terribly stupid to the experienced builders who were working under my direction. Between the books and tips from the others I learned a whole lot of carpentry over the next eight weeks.

One fellow, a speed-freak named Buddy, was particularly helpful, his mouth moving as fast as his hands, and happy to share his broad experience while he worked. He was living on coffee, caffeinated cola and black beauties [a then-popular form of amphetamines] and interspersed his thoughts about chalk lines and nail guns and stud spacing and rafter layouts with tortured fretting about his girlfriend back in California who he just knew was running around on him. He was trying to decide if he should take her back or strangle her.

We worked fifteen or sixteen hours each day, taking advantage of the long summer sunlight in that near-Arctic region and collapsed into the back of our VW Squareback each night, too weary to talk. I'd read a few pages of building books by flashlight and fade.

Carpentry work in Alaska was a spoiler. The quality of the wood—at least as it was three decades back—was unsurpassed. The Douglas fir and cedar lumber was sawn from old growth giants, the grain was clear and the wood came in any dimension needed. Most of the lumber I have handled in the years since has seemed a little shabby and knotty and poor by comparison. The big forests are disappearing, and while I have surely had a hand in the deforestation I recognize that my life is the richer for that experience. I have worked with some of the best timber ever harvested on this planet and it is unlikely that it will ever be seen again.

Late September brought the first sign of snow on the peaks. "Termination dust," as it's known in Alaska, is the warning that the Al-Can highway will soon be impassable in its higher reaches, and so the time arrives to decide to stay for the winter or head south. We collected our pay, loaded the canoe atop the Volkswagen and started for New Hampshire.

Car camping in a squareback VW is a little tight. At night we'd put the cooler and other gear on top of the canoe on the roof of the car to make room for sleeping. With the front seats kicked forward there was barely enough room for me to stretch out my 6'3" frame, but it was a whole lot quicker (and drier) than setting up a tent at every stop along the way.

One night we were awakened by growling. Snook, our four-year-old calico cat was perched on the dashboard staring out the windshield and making menacing noises. There was a mid-size black bear in front of the car and just as we awoke, it started to climb on the hood, obviously headed for the cooler. I hit the horn, but the hungry bear continued, clambering upward, now standing on the hood and clawing at the canoe. Snook kept snarling. I pulled myself over the seat, put the key in the ignition and started the car. The bear slowly climbed down, walked about ten feet and sat down on its rump, staring at us.

I gunned the engine and it stood again then walked away, looking back a few times as if reconsidering its retreat.

A few hours later Snook started up growling again, and sure enough our bear had returned. But this time it just sat and looked and soon continued on its way.

Days later we pulled into a trading post that featured two life-sized, rampant grizzly statues out front, made of painted wood or plaster. Snook took her position on the dash again and growled and hissed the whole time we were stopped, apparently certain of her ability to scare off the giants that threatened.

We had earned enough money that we decided we could afford to take the ferry back south, rather than drive the whole Al-Can again, and besides, cruising the inland passage seemed like a great adventure. So we veered south to Skagway where we learned that ticket prices would drop substantially on October first, making it well worth a few days delay. While we waited we chanced to hear a radio request for help from the wildlife authorities, and decided to volunteer a day's work.

The Skagway River had flooded and receded, leaving a sandbar that trapped thousands of migrating salmon in a large, temporary pond. The project was to catch the milling fish, put them in plastic trash cans, haul them over the sandbar and release them in the river. The rangers provided chest-high waders and we plunged into the water with dozens of other volunteers, working all day to free countless huge fish, some close to four feet long.

Many salmon were already beached and dead or dying and bald eagles were ripping into them. Literally hundreds of eagles were perched atop surrounding trees, waiting their turn or waiting for we interlopers to leave. It was an astonishing scene. By the end of the day there were far fewer fish swirling in the muddy pond and we were all flagging from the chill and the effort. I seem to recall a hot shower at the end, but I can't quite remember how that happened, since we were camped in the woods. Maybe its retroactive wishful thinking.

A couple of days later we boarded the boat.

That was almost a decade before the *Exxon Valdez* fouled the shores of Prince Rupert Sound and the scenery was spectacular. I spent most of each day on the deck watching the forest, the glaciers, the fishing boats and the sea birds.

When the *Valdez* ruptured and disgorged its viscous cargo in Prince William Sound, dead animals, idle fishing boats, briefed lawyers, de-briefed sailors, disquieting headlines, accusations and counter-accusations followed. Altogether it was a royal mess and I called up vivid memories of pristine, forested vistas broken by glaciers tumbling into the sea. The coastline affected by that spill is as long as the Atlantic shore from New York to Miami, a fact obscured by our standard

Mercator projected maps. And while that spill was raised to high relief by stories of the captain's malfeasance and Exxon's continuing refusal to pay for the damage, it doesn't come close to being the biggest oil spill we've visited on our oceans. Not by a long shot.

I once discovered a beached seal, suffocated in oil on the Olympic coast. Its eyes, nose, mouth, fur, fins and tail were coated in a thick black sludge. One dead animal is a graspable concept which I might extrapolate to thousands while viewing pictures from the Alaska debacle —though the numbers quickly blur. In the same way, in the other Washington, I see the face of that boy in my high school class when I gaze down the length of a black marble wall, the particular Vietnam casualty set against the horrific reality. I know but I don't know. Perhaps I can't.

Somewhere south of Haines, I was leaning on the rail savoring the view when a humpback whale breached. The animal cleared the water but for its tail, its white fins and huge mass seeming to hang suspended before it arched and crashed back into the water.

Adult humpbacks range in length from forty to just over fifty feet and weigh up to forty tons. The humpback has a distinctive body shape, with unusually long pectoral fins and a knobbly head. "My" whale's species was instantly recognizable. In addition to their proclivity for acrobatic breaching, humpbacks are known for the male's complex whale song, which lasts for 10 to 20 minutes and is repeated for hours at a time. Experts say the purpose of the song is not understood, although it appears to have a role in mating.

Do any complex songs from humans *not* involve mating?

Humpbacks migrate up to sixteen thousand miles each year. They feed only in summer, in polar waters, and migrate to tropical or sub-tropical waters to breed and give birth in the winter. So my whale was starting south just as we were. Come winter, humpbacks fast and live off their fat reserves, not unlike any number of people I've met in the construction trades who draw unemployment during the winter months.

Like the right whale, the humpback is a baleen whale, and was hunted mercilessly until its original numbers were reduced by 90 percent or more. It was deemed to be on the brink of extinction when a near-global moratorium took effect in 1966. Today it is believed that there are at least forty thousand humpback whales worldwide but in 1981 my whale was probably one of only about ten thousand in all the world's waters, since Soviet whaling had continued after the ban well into the 1970s. Numbers only began to recover about the time I saw mine go airborne.

Susan, who was inside the cabin reading when the cetacean acrobat leapt, was immediately pissed. "Why didn't you call me?" she demanded when I hurried inside to tell her what I'd seen.

"That's why I came in."

"Asshole," she said. "Thanks a lot."

I spent even more time on deck for the remainder of the journey, braving pretty consistent and decidedly chilly rain, hoping to see a recurrence, but no luck.

5. hemlock

In the 1970s we fought back.

We were among the many New Englanders opposed to construction of the Seabrook Nuclear Power Plant and so joined the anti-nuclear Clamshell Alliance which had formed in Portsmouth, New Hampshire. We participated in rallies and demonstrations, though neither of us chose to join the May 1, 1977, occupation of the site, which resulted in more than 1,400 arrests. But soon thereafter a move by Public Service Company galvanized me to take action.

With government blessing, Public Service tacked on a Construction-Work-In-Progress surcharge to all of its customers' electric bills, starting in 1978. That is, we were suddenly expected to pay for construction of a privately owned nuclear power facility whose output would be sold out of state and across the border into Canada. New Hampshire didn't need the power. That was the final straw as far as I was concerned, and after lengthy discussion in the winter of 1979 we decided we should relocate to Ontario, Canada where some friends owned a large swath of land.

On March 23, Susan took off to spend a week with a few of her friends in a cabin somewhere in eastern Pennsylvania or New Jersey. She was typically vague about her plans but assured me that there would be no phone and that she would be home the following weekend. Five days later Unit Two in a nuclear facility known as Three Mile Island near Harrisburg experienced a partial core meltdown and news reports were inconclusive, contradictory and run through with misinformation from government officials and agencies.

Pennsylvania's Lieutenant Governor, William Scranton, told the press there was nothing to worry about, then a day later indicated that the accident was worse than previously reported. Conflicting numbers concerning radiation releases were offered. Schools were closed, people were urged to stay indoors and farmers told to keep livestock under roofs and to feed the animals only stored food. Nuclear Regulatory Commission Chair Joseph Hendrie advised Governor Dick Thornburgh to encourage the evacuation "of pregnant women and pre-school age children ... within a five-mile radius of the Three Mile Island facility." A frantic exodus began, jamming highways as one hundred and forty thousand people left the area.

Abetting the widespread sense of panic that pervaded the nation was the release on March 16 of a Hollywood thriller film, *The China Syndrome*, which centered on the possibility of a catastrophic nuclear meltdown. Abetting my personal sense of panic was the fact that Susan was somewhere downwind of Harrisburg and could not be reached by phone. I slept very little over the next few nights and disconsolately played my guitar between listening to news updates and hoping for a phone call. I wrote a song that summed up my thoughts.

> *I'm sitting in uncertainty tonight,*
> *A lonely fire before me, my thoughts in disarray.*
> *Like the fading failing of the light as evening tumbles*
> *down, my soul in winter shadows wonders on.*
> *Shallow wells run dry in time and summer flowers fade,*
> *There's so much more I need to feel, before I feel my age.*
> *Woman when you're gone,*
> *my feelings come so clear to me.*

Woman when you're gone,
I know I need you near to me.
Oh Lord, when you're gone, life loses continuity.
Oh my woman.

I'm sitting in uncertainty tonight,
A lonely fire before me, my thoughts in disarray.
Like the fading failing of the light as evening tumbles
down, my soul in winter shadows wonders on.
Like the fading failing of the light as evening tumbles
down, my soul in winter shadows wonders on.

Susan showed up on Monday and blew off my expressed concern. "Yeah, I heard about the nuke. Big deal."

"Why didn't you phone? Didn't you think I'd be worried?"

"We didn't have a phone and we weren't anywhere near Harrisburg. Why would I think you were worried?" Matter closed.

Later that evening I played the song I'd written for her during my vigil.

"That's pretty stupid," was her only response.

We continued to prepare for our pending move to Ontario. However, a friend talked us out of that scheme by offering a more-than-fair deal on eight acres of his property in Deerfield, and in spring of 1979 we set about constructing a cabin in a place I christened Mountainview Swamp—located as it was on Mountain View Road, and being half wetland and marsh.

Building that cabin would constitute the better part of my carpentry experience when I sought work in Alaska two years later, and it partook more of pioneer homesteading than plumb-square-level craftsmanship.

I bought a small chainsaw and cut and limbed medium sized hemlock trees that abounded on the property. I cut the business-end off a partly broken shovel so the remaining blade was about six inches long and four wide, and sharpened the leading edge to make what's called a spud peeler. Using that tool we peeled the bark, still loose with the rise of spring sap. It came off pretty easily in long strips. The construction style was "pole-barn," a sort of post-and-beam construction with round logs, consisting of uprights the full height of the building, cross-members to tie the uprights and support floor joists and rafters, and angle braces to keep the resulting box from shifting.

The project involved heavy physical labor, dragging and peeling logs, digging foundation holes then lined with field stone, upending the poles and driving 30-penny spikes (these are 4 inch nails as thick as a pencil, and this job came before my lesson about bigger hammers). We hefted peeled-pole rafters into place, all the while swatting mosquitoes and black flies intent on entirely draining our circulatory systems.

The wall, roof and floor surfaces were fashioned from slabs obtained free from a local saw mill. (When a log first passes through a mill the rounded first cut "boards" are discarded or chipped into mulch.) Roof shingles and heavy-gauge tar paper for the exterior siding came from a salvage yard. Windows were discards from a remodeling job site and the front door set us back $5 at a yard sale. The sink was salvaged as well and plumbed from a tank on the outside of the building that was filled with a bucket.

Electrical wiring was free from a local dump, as were outlets and light fixtures, and we ran lights and radio with a car battery. We had successfully unplugged from Public Service Company and a year or so later we upgraded our power source to photovoltaic panels. I wouldn't have any cause to return to grid power (at home) until the year 2000 more than 20 years later.

In the end, the 160 square foot cabin with an 80 square foot sleeping loft cost a grand total of $97.

Addition of a porch and a chimney in later years would drive the total investment to almost $120 (or fifty cents a square foot) and the building served as a three-season dwelling for the next two decades. It may still be standing today, though I haven't seen it since 2002 and new owners may have considered it more an eyesore than a useful utility building. A friend even overwintered there once, though I have to guess it was grindingly cold in a place where February nights are often sub-zero.

Our generous neighbor graced us with an out-house, and insofar as he was a cabinet-maker, the outhouse was in every detail vastly superior to our make-do cabin. In due course the cabin became known as the Castle and the outhouse, naturally enough, as the Throne Room.

Some years later we swapped firewood for an elegant used chicken coop, also built by a cabinet-maker who delivered it on the trailer he used to haul his wood home. We muscled the building downslope on skids into the midst of the garden and set it on a dry-stone foundation. Scrubbed and lined with cedar scraps from a job site, with a claw-foot tub and small wood heater, the coop became a bath-house and sauna, whence a bather could gaze out a large circular window at flowers and veggies in summer, or snowy hemlocks when the season changed.

We had spent the summer building the cabin in between doing masonry jobs—chimneys, fireplaces, foundations—and when the snow began to fly we were ready for a break.

"We're going to Tucson," Susan announced.

"Tucson?"

6. jonah

Susan owned a Supervan.

She and her husband had outfitted the 1968 Ford truck as a camper with built-in cabinets, a bed and dining table, a roof rack with storage compartment and room for the canoe. The interior shelving, floor, walls and roof unit were built with 3/4 inch marine plywood, assembled with stainless screws and fitted out with indoor/outdoor carpet. The thing was built like a tank and weighed half again what a stripped vehicle would run. It was solid and quiet and comfortable and had been their home for a year or more of travel around the country. Somewhere along the way it became known as The Mother Van.

In the summer of 1977 The Mother Van spun a bearing. Fortunately we were renting a century-old farmhouse with a big flat gravel parking area perfect for pulling engines and other intravehicular frivolity. The spread included a big garden space, ducks and a pond. The ducks belonged to the home-owners and part of our rental agreement included duck feeding—the food consisting of several crates of army surplus cookie-cracker rations in olive drab tins. Vietnam War leftovers, I suppose.

By this point I had already been T-boned and totaled in my orange VW Beetle in a snowstorm (no fault, no foul, no insurance payout) and then traded my incredibly beautiful Gibson Les Paul low-impedance electric guitar for a Toyota Stout pick-up. The Stout was an early Toyota contraption which pretty thoroughly sucked. Bad move, I still wish I had kept the guitar and found some other way to buy a truck, a *different* truck ... but stupid mistakes are free.

At least we had a second vehicle, and a truck to boot. So we rented an engine hoist, bought a Chilton's Guide and pulled The Mother Van's motor up and out via the passenger-side door.

Taking an engine apart is the best way to understand how engines are put together—in the same way that dissecting a frog or disassembling a laptop computer engender insight into structure. We did a pretty decent job, for beginners, with only a small handful of parts left over (internal combustion engines are *way* more forgiving than computers on that score).

Theretofore my mechanical experience had been limited to peripheral matters: water pumps, alternators, clutches, brakes, U-joints, tune-ups and the like. The main thing is to keep everything organized and clean. A machine shop turned the crank shaft (that is, ground it down so the bearing surfaces were true and smooth again), bored the cylinders and did a valve job on the head. Then we bought new bearings, rings, valve lifters and gaskets, torqued it all together and planted it back on the engine mounts. It ran so well we decided The Mother Van deserved a paint job too, and a friend with a spray rig helped me paint it a lovely shade of maroon.

That first year in the farmhouse is memorable for many reasons beyond zen and the art of mothervan maintenance, and though the flying saucer incident and the airborne giant ice cube from hell undoubtedly made the second year a little weirder, it is 1977 that mattered more in the long run. To wit:
  • *I read R. Buckminster Fuller's magnum opus,*
  Synergetics, *cover-to-cover for the first time,*

understood some of it, and was totally blown away.
• Snook, the calico cat, had nine kittens, and I kept the runt, naming him Brave Ulysses to give him courage. He was the best cat I ever shared my life with—except for the time he pissed on me from inside a backpack which wasn't entirely his fault—and he was always extraordinarily brave except for that night we saw a lynx near the campsite, in Alaska. A gun-toting bastard shot him dead when he was only ten years old.
• A 4x4 driver intentionally flattened Cochica.
• A gun-toting bastard shot Tom.
• A bullfrog tried to swallow a duckling.

Yes, that's right. One morning there arose a calamitous tumult out at the duck pond with the adults squawking and honking and flapping and the dozen or more ducklings squeaking and dashing in every direction.

I ran out to the pond to discover a frog trying to eat a duckling about its own size, perhaps mistaking the bird's head for a large bug. The frog couldn't swallow the bird and the bird couldn't extricate its head. It was flapping its tiny yellow wings and kicking it's feet, while the frog kicked in the other direction, gap-mouthed and wide-eyed and unable (unwilling?) to spit out its prize.

I picked the pair up and slowly drew the duckling out, which pulled much of the frog's esophagus inside-out as well. The duck went scooting back to join its nestmates and parents while I stuffed the frog's innards back down its throat. It composed itself a moment on the bank, then jumped into the pond and swam away.

The frog became our stand-in for "a great sea creature" and the duckling was henceforth and obviously known as Jonah.

During his days inside the great fish—or whale, depending on your preferred version of the tale—the Biblical Jonah repents and obeys the call to prophesy against Nineveh, the Ninevites repent in turn and God forgives them for being recalcitrant Ninnies.

Our Jonah may or may not have repented, but he prophesied pretty well. His neck was twisted a little toward the right, and permanently bent, and when his adult plumage came in he had a wild feather that grew upward on that side of his head. It made him look like he was wearing a kid's costume Indian headband and he tended to walk very fast, often in circles and squawk nonstop. So as the flock would cross the yard from the pond toward the house where we scattered K-rations, Jonah would initially race ahead but curve back into the gang, preaching like mad, then curve some more and race to catch up.

Whether Jonah's flock heeded the call remains somewhat a mystery. But with unfenced land the birds had an unfortunate tendency to traipse into the road and most of them were ultimately hit by cars. Jonah was a survivor, avoiding the traffic through dumb luck or, who knows, foresight?

The Jonah story offers one of those odd tests of faith that populate religious faith. No one with any understanding of physiology could imagine that a human could survive three days in the digestive tract of a beast, and of course that's the point—it could only have been a miracle that permitted the prophet a chance to reconsider his wayward path and be delivered intact to a beach. But, then, whose word do we have on that score? Why, the word of a raving prophet who claimed to have been swallowed and spit up on a spit.

Seems, um, a little fishy to me.

When I came to write a biography of the evangelist Billy Graham many years later I was startled to learn that while he had rejected a literal interpretation of the Biblical creation story and allowed that the seven days could be read as an allegory for the billions of years over which our solar system and planet evolved into existence,

50 Cecil Bothwell

Graham in his penultimate decade still held to his belief that Jonah had actually spent three days inside a whale.

Embracing faith seems ultimately a matter of choosing to believe in the putative benefits of having faith, and where one chooses to draw one's lines seems as arbitrary as taste in music or art or literature. We seem to say, "This set of rules speaks to my need for mystery or inspiration. That set does not."

Of course, the faithful generally assure others that lack of faith in their version of reality will result in an eternal journey through hell, or reincarnation for another shot at enlightenment, or conquest by heathen enemies, or crop failure, or bewitchment, or disappearance of the sun from the sky ... there are warnings aplenty for the apostate. As I was editing this passage, the Christian talk-show poobah Pat Robertson announced that the 2010 Haitian earthquake is punishment for a supposed pact with the devil made by the denizens of that sad nation—just as he previously opined that Hurricane Katrina was God's vengeance on sinful New Orleans and 9/11 God's reaction to sodomites in the Big Apple. Okay, Pat. Whatever.

In the early days of whaling, Nantucketers are said to have upgraded their techniques in 1690 with the expertise of a Cape Codder named Ichabod Paddock. Paddock's claim to fame was to have been swallowed by a whale in whose belly he found a mermaid and the Devil playing cards for his soul. What Nantucketers thought of their instructor's elaborate history is less than clear, but his hunting expertise was deemed a boon to industry.

Cochica the retriever cat, faithful as the day was long, was flattened by a truck along that New Hampshire by-way too, just like the unrepentant ducks. But she was asleep in her usual nest about fifteen feet from the pavement and telltale tread marks made it pretty clear that someone swerved a long way to do the deed. Some people don't believe in living cats.

Toward winter we adopted (or had foisted on us) our neighbor's cats—Moxie, a fluffy red tabby, and a grey short-hair tabby, decidedly outdoor, feline named Tom. He was scruffy from numerous cat fights in the way that unneutered toms usually are, with a tattered ear and missing tufts of fur. But he was quite sweet to people and purred like mad when anyone found time to scratch his head.

Tom showed up one evening, dragging a leg, the bone shattered by a bullet that was still lodged in his chest. The vet said he might recover after amputation and surgery but would need to be kept indoors for months as he mended and learned to walk on three legs. Given that he had no experience with use of a litter box and was a persistent scent-sprayer we couldn't see our way clear to taking on his troubles. We had him euthanized.

That decision set us up for a far more painful experience just a few months later when, worn down by the New England winter, we loaded The Mother Van and headed south to visit family and friends in Florida.

7. peace

Florida's flatwater rivers are idyllic.

They offer some of the finest and most picturesque canoe runs on the continent and, in the cold months, a wide variety of overwintering birds lend excursions a palpable exoticism. You round a bend and a huge flock of white ibis fly overhead, close enough that you feel the rhythmic pulse of air beneath their wings. They tilt in the sunlight alternately flashing bright and dark like a flickering school of reef fish. Unseen limpkins laugh like George-of-the-Jungle-talking-tiki-birds as they hunt snails in the rushes while a pair of bald eagles circle high above. Eagles dive-bomb ospreys to steal fish—the fish hawks dropping their prey in fear and the eagles swooping below to snatch the prize in mid-air. Four varieties of heron and magnificent wood storks whistle and croak in the marshes and stately sand hill cranes strut along the banks.

By late winter alligators have come back up from their muddy hibernation and bask in the sun on the banks, some slithering below the surface at one's first approach, others exploding into motion and thrashing their way into the river's safe haven. A 10-foot gator erupting a canoe's length distant creates a very impressive departure and even after decades of exposure to those giant reptiles I still experience an electric thrill of fear and joy in the presence of their unmediated wildness.

It's the scene I always imagine when I hear Thoreau's phrase, "In Wildness is the preservation of the world."

That season we decided to explore the Peace River which originates about sixty miles southwest of Orlando and flows for more than one hundred miles to where it empties into the Charlotte Harbor estuary on the Gulf coast. We started at the south end, drift fishing in the harbor over a few days, then paddled up into the delta where hummocky islands divided the several channels.

A pod of dolphins appeared one afternoon, driving a school of fish into the shallows around our canoe, a couple of their number eyeing us warily while most cavorted, enjoyed their fish feast and seemed to completely ignore our presence.

Fresh water from the Peace River is vital to maintain the delicate salinity of Charlotte Harbor that hosts several endangered species, as well as commercial and recreational harvests of shrimp, crabs, and fish. The dolphins were certainly having a blast in their fishery, their sleek, grey bodies whirling and glistening as they crested the surface and charged amidst the scattering prey.

Early Anglo explorers discovered scads of pleistocene and miocene fossils throughout the Peace River basin, which led to the discovery of phosphate deposits. Subsequent phosphate mining and processing emitted vast quantities of fluoride which would eventually poison hundreds of thousands of acres of Florida farmland in an area picturesquely referred to as "the Bone Valley." Dem bones, dem bones, dem dry bones.

Among those 20-million-year-old fossils were remnants of the earliest toothed whales, forbears of the bottle-nosed dolphins that swarmed around our canoe and the orcas we would spot two decades later off the coast of Newfoundland. Though most toothed whales are smallish, sperm whales are members of the clan as well, and constituted a major target for the whaling industry. Unlike their baleen cousins, which were hunted primarily for fuel and "plastic," these giants were initially slaughtered to make candles, soap, perfume, machine oil, leather waterproofing, rustproofing, pharmaceuticals and crayons.

One of my earliest extant childhood drawings is a crayon depiction of a breaching sperm whale. That's an oddly ironic memory viewed through this later-on lens, though by the 1950s Crayolas were petroleum- or tallow-based, so there is no possibility that I was drawing my whale with whale grease.

Sperm whaling on a commercial scale began around 1700 and increased until the mid-1800s. Sperm-aceti oil, cleaner burning than other whale oils, was preferred for lighthouse lamps and became important in public lighting before 1862 when it was replaced first by lard oil, and then petroleum. It was also a major lubricant for the machinery of the Industrial Revolution.

It's pretty clear that sperm and other great whales were saved from utter extinction by the rise of the petroleum industry. Oil giveth and oil taketh away. But beef cattle were part of the process as well. In the mid-1800s beef tallow was cheaper than spermaceti for candles, and lard oil was outcompeting whale oil for lamps. About the same time time a Canadian figured out how to create kerosene from coal, which burned cleaner than either lard or whale fat. Soon Pennsylvania petroleum kicked in and commercial whaling in the U.S. became obsolete: fossil fuels were cheaper than blubber.

The early toothed whales whose fossils are found in the phosphate deposits gradually gave way to the species we see today. Scientists speculate that the most likely reason for the demise of the earlier forms was global climate change circa 14,000,000 B.C.E.

Climate change matters.

Michael Connett, reporting for the Fluoride Action Network, has documented that phosphate mining and processing has been a major source of fluoride pollution in the Peace River region. He wrote:

The source of the problem lies in the fact that raw phosphate ore contains high concentrations of fluoride, usually between 20,000 to 40,000 parts per million (equivalent to 2-4 percent of the ore). When this ore is processed into water-soluble phosphate (via the addition of sulfuric acid), the fluoride content of the ore is vaporized into the air, forming highly toxic gaseous compounds (hydrogen fluoride and silicon tetrafluoride).

In the past, when the industry had little, if any, pollution control, the fluoride gases were frequently emitted in large volumes into surrounding communities, causing serious environmental damage.

Connett goes on to note that in Polk County, Florida, the headwaters of the Peace River, the creation of multiple phosphate plants in the 1940s caused damage to nearly 25,000 acres of citrus groves and "mass fluoride poisoning" of cattle. According to the former president of the Polk County Cattlemen's Association:

Around 1953 we noticed a change in our cattle ... We watched our cattle become gaunt and starved, their legs became deformed; they lost their teeth. Reproduction fell off and when a cow did have a calf, it was also affected by this malady or was a stillborn.

The river we explored that winter was called Rio de la Paz (River of Peace) on 16th century Spanish charts and it appeared as Peas Creek or Pease Creek on later maps. Somewhat confusingly the native Creek and Seminole Indians call it Talakchopcohatchee, River of

Long Peas. This was presumably before Andrew Jackson and company disturbed their long peace with butchery and near extermination.

Several days further on we paddled a stretch of the Peace near Wauchula. Typically, we had parked The Mother Van in deep shade in a densely wooded area, to keep the cats comfortable, and upon our return close to dusk, we let them out for a romp. Snook, Brave Ulysses, Moxie and Quinta (a.k.a. Big Mama) hopped to the sandy ground and sniffed around cautiously before stretching against tree trunks and then spreading out to explore the new campsite. They would come and go over the next hour or so, as was their custom, before we'd call them in and feed them.

A shout broke the still evening, "Look at that fucking cat!" Then a gun shot.

Snook, Ulysses and Moxie were back in the van in a heartbeat. I ran down to the river and spotted two men in a john boat, fishing rods in hand and a shotgun barrel poking up over one gunwale. I recognized the boat, having seen it in the bed of a pick-up parked a short distance upriver earlier in the day. In the course of searching for Quinta I went back up to the truck and wrote down the tag number.

We called for hours, without luck. Then, close to dawn, we were awakened by a pitiful mewing and scratching. Fifteen year old Quinta was trying to get into the van. Much of one rear leg was missing and the rest was a bloody mess. We cleaned her up and calmed her down, and couldn't make morning come fast enough.

At 9 a.m. we took Quinta to the nearest veterinarian we could locate, in Wauchula. He said he believed she could survive if he amputated her leg and stitched her up. An X-ray showed no other broken bones or peripheral damage outside of some embedded shotgun pellets. So, fully burdened by our decision concerning Tom, we assented to the surgery.

"Their legs became deformed ..."

Meanwhile we visited the sheriff's office and reported the shooting. The deputy smirked and informed us that the shooter was the Sheriff's nephew and that we were wasting our time since the Sheriff probably shot cats on sight as well. Susan managed to get the name of the shooter and by the end of the day we located the house and she was engaged in a shouting contest with the young man and his father.

Finally the father went into the house and came out with a shotgun to tell us to leave while we could still walk. Sheriff, deputy, nephew, father: four more names on Susan's list.

Three days later the vet informed us that Quinta was healthy enough to be released. We settled a large bill and carried her to The Mother Van. She seemed groggy and pretty listless. Fifteen miles down the road Quinta suddenly let out a blood-curdling scream and dropped dead.

" *... or was a stillborn,* " as the cattleman said.

8. whirl

I am not a real birder.

I've kept no "life list" of myriad species observed. There have been no journeys around the world to spy endangered rarities. And I am terrible at remembering calls—barring a few blatantly distinctive exceptions, most of the twitters, whistles and clicks that birds utter leave me clueless about their identity.

But I love to watch, floating downriver in a canoe—beer in one hand, binoculars in the other, steering-paddle tucked under my arm—keenly alert to the wild things all about.

There's a red-shouldered hawk perched atop a dead palm trunk, tearing at the unseen prey pinned in its claws. Chunks of wet, red something, ripped and swallowed. A white-spotted brown limpkin wails like a banshee, then resumes poking its long, downturned beak between weeds, in search of apple snails. Now a dull grey eastern phoebe zooms overhead, stalls for two beats and veers to a nearby branch, its twitching breakfast bug briefly visible. And gone.

Unless you spot birds engaged in courtship or building nests, their primary interesting activity is eating. Birdseed is big business, no? Not so different from us, really: We woo. We find housing. We eat. And, having our three squares passably covered in the winter of 1997, Susan and I were back in the canoe again, watching.

We were paddling the Ocklawaha River in the northern portion of the Ocala National Forest in central Florida. This particular journey was more than usually colored by death—always there, to be sure, but this time not easily ignored. An American alligator floating upside down in the weeds, and a river otter similarly positioned against a drifting log, attested to someone's able, if misguided, marksmanship. A young grey fox showing no apparent fatal wound lay stiff amid pine needles, which still bore the imprint of its final contortions. And, strangest of all, beneath four feet of clear, flowing water, a white-tailed doe.

We paddled back over her, waited for ripples to flatten, and looked again. Completely intact: Two forelegs hooked around a submerged branch kicked languidly in the current. Her head was thrown back, and she lay belly up, little teats attesting to her gender. We guessed she had not been dead long, else the gators and turtles would have parted her out.

But there were other shadows, too: Each of us had a mother's sister enduring untender cancer therapy. We are all terminal. Some of us are led to believe that we've been more specifically informed, but no matter. We each owe the earth one body—our tuppence for the piper who has favored us with this lovely tune, this wondrous dance.

Soon enough both aunts would succumb amid whispered goodbyes and morphine-induced dreams.

Geologist Vladimir Ivanovitch Vernadsky referred to life as a "disperse of rock." We are not separate from the earth's crust, Vernadsky observed, we are just the parts that visibly wiggle. Agitated molecules whirling into dervishes, CEOs, peasant farmers and canoe paddlers. Moving, always moving.

And what motion! Vultures, both black and turkey, soared together overhead. Up and up into the yonder, till binoculars were insufficient to follow their flight. Searching for leftovers and riding the wind. Nice work if you can get it—perfect players in the recycling loop, probate jurists in this mineral disperse.

Apart from the buzzards (and headed higher), an adult bald eagle, wings stiff as twin ironing boards, circled above a juvenile baldy practicing her moves. I once met a woman who hang-glides for sport; she recalled rising impossibly high on an updraft above Mount Shasta—wingtip to wingtip with our national bird. She said the eagle first seemed curious about the Dacron-fledged interloper in its airspace, but presently grew bored and moved on. Up.

I looked up. There, atop a bleached pine snag, sat a vigilant osprey, black talons biting into pale wood. I aimed my binoculars and was startled by the view. Above the feathered shoulder hung the waxing, gibbous moon, a semicircular mirage set in aching blue. Wondering under my gaze, the bird's head pivoted, and together, we considered the familiar lunar image.

The osprey turned back with a shrug. "Oh, that." Nothing there but dust. No water there for dancers to drink. The party never even got started.

Here on terra firma, though, the party never ends. I ate another boiled peanut, sipped my beer, and watched four eastern painted sliders plop off a log. Miracle after miracle, we celebrate, we whirl.

Three years later it was Susan's turn to confront a cancer diagnosis, and our world turned upside down.

"Don't count on tomorrow," it said. "Do it today."

Death is necessary to life, it seems. In his exquisite treatment of life in *Sex* (Science Writers Books, 2009), Dorion Sagan quotes the character Hannibal Lecter (*The Silence of the Lambs*), "As a sexual being, Agent Starling, 'we' must die."

Sagan observes that we exist in a Red Queen's race, where, like the Queen and Alice running in Wonderland we have to constantly evolve to maintain our status quo.

Sexual reproduction lets parents shuffle the genetic deck creating small variations that (with luck) inure offspring to the ever-evolving parasites, viruses and bacteria nipping at our collective heels. Then the parents succumb, and the cycle repeats.

In the time between we celebrate and whirl.

9. bald

We were perched on a crumbling sea wall.

It was late January, 1981, we had just arrived in Mazatlan, Mexico, and I savored the warm sea breeze wafting across my newly shaved head. The global change in one's relationship to the physical world that attends abrupt removal of scalp hair is singular.

I highly recommend it.

The spray of a shower or the splatter of rain suddenly touch nerves that have been shielded for years (in my case, thirty of them). The warmth of sunshine and overhead lights seems to radiate inward from the now exposed skin. Diving into a swimming pool or porpoising into a breaking wave becomes a wholly new experience, as does the touch of a cool pillowcase, or, as in this instance, a warm breeze.

In that moment in Mazatlan my newly hairless dome led to a singular conversation with a garrulous retiree, a discussion which would trigger nerves inside my head with lasting effect.

The shaving itself had been accomplished as a lark, somewhat abetted by tequila. Susan and I had finished a year's stint as house parents for developmentally disabled youngsters and we had returned to Tucson to celebrate before launching on our next adventure.

The Tucson trip the previous year had started with a very brief Thanksgiving layover in Canton, Ohio, where

Susan's mother had yet to come to terms with our cohabitation. In later years she would reach an agreement with God that our sin was between us and Him, and permitted us to stay at her home during extended holiday visits. But that lay in the future, so we only stopped long enough to eat turkey with the family and pick up her brother's bass guitar amp, delivery of which had turned out to be the motivating rationale for Susan's choice of destination.

The late November drive from Canton to Tucson included the most surreal highway scene in my memory—an ice storm on Interstate 30 between Little Rock and Texarkana. We were headed up a long grade on a slush-covered highway with sleet pounding on the windshield that reduced visibility to near zero. Our headlights simply lit up a white wall of precipitation ahead as we moved forward at a few miles per hour. Beside us were dozens of tractor-trailer rigs, wheels slowly spinning in place, wobbling back and forth but not advancing at all on the icy pavement. At times the falling ice would blow forward faster than we moved forward, lending the sense that we were actually rolling backward, or simply adrift and floating. There was no way to see off-ramps, or even to change lanes, so we just kept moving.

Besides which, the amp and speaker case pretty much filled the sleeping space in The Mother Van, so we were determined to drive straight through, and finally broke out of the foul weather the next day, a good way west of Abilene. It was a very tense 24 hours.

We landed in Tucson and spent the holiday season with Susan's brother and housemates just west of the city. We hiked up buttes and along arroyos, visited the Sonoran Desert Museum to learn about the local fauna and flora and savored the warmth.

One morning Susan announced that she'd found us a job and that we were going to be house-parents for mildly developmentally disabled/mentally retarded kids. For a year.

Well, whatever.

We interviewed and were deemed acceptable. Our lack of experience or education in the field was no obstacle because the company we were going to work for was in desperate need of worker bees. The state of Arizona had joined the national trend toward de-populating mental institutions, moving clients into group homes and normalizing their lives as much as possible. The program was in full swing, grants had been granted and the game was on. We were trained for a few hours including a very short course in a simplified version of American Sign Language, and given some handouts. During the second week of January we moved into a house on South Palomar Drive with six severely DDMR clients from age 12 to 20, and one 14-year-old girl who was wildly psychotic and whose eyes seemed to track separately.

We were on duty from 4 p.m. Sunday to 8 a.m. Friday, with relief parents spelling us in the gap. We drove the kids to school each weekday by 8 a.m. and picked them up in the afternoon. They were being "mainstreamed" which meant that they were attending regular public schools, though I never inquired too deeply into how the schools were handling the load. Our crew would have been impossible to deal with in any kind of normal educational setting, and while I understood the rationale for normalizing their lives—and grasped the social and political intent of exposing the general population to differently abled people in order to build public understanding—they had to be a tremendous burden for the teaching staff.

In the morning we would walk each child through the steps of getting dressed, making the bed, brushing teeth, and taking their portion of prescription medications. (That is to say, we would dress them, make their beds, brush their teeth and toss pills down their throats.) After school hours we were tasked with improving our client's language and living skills. One of us would spend a half hour with each of them in turn. (Note: I've changed the names to protect client privacy.)

You help the child form his hand into a fist, and wiggle it like a sock puppet nodding its head.

"Yes, Donny, this means 'yes.' Can you say 'yes'?"

Donny unclenches his hand as soon as you release it and flings his glasses at the wall while biting his finger and shouting. "Whawunh! Wha! Whuh!"

You fetch the glasses, put them back on his nose, form his hand into a nodding fist.

"Yes, Donny, this means 'yes.' Can you say 'yes'?"

Repeat for half an hour, then start up with Jim.

"Yes, Jim, this means 'yes.' Can you say 'yes'?"

Jim mumbles a kind of simulacrum of "Thank you," sort of a "anh oo" and makes the appropriate gesture—palm flat toward the chin, with fingers touching the lips, then moving the hand down and away. Then an unrelated motion, bringing the back of his hand close to his glasses and staring at the fingers as if in surprise.

"You're welcome, Jim," you answer, trying to keep your response appropriate. Then you form his hand into a nodding fist.

"Yes, Jim, this means 'yes.' Can you say 'yes'?"

"Anh oo," he repeats, accompanied by his sole sign.

Repeat for half an hour, then start up with Terry, who shakes his hand out of yours every time you attempt to help him form a fist, flails about and makes weird vowel sounds while pointing at the wall and nodding vigorously. "Aaaanh. Aaaanh. Uuuuuuh."

While I was busy with the boys, Susan would be in the next room going through the same motions with the girls, differing only in that her sessions with Trina were actual conversations, of a sort.

"Lisa says she wants to go outside now."

"Trina, we can go outside after we work on the lesson. Can you read that first paragraph?"

Trina rips the book in half. "Lisa did it. Lisa did it! She wants to go outside!"

Susan locates the first page and places it in front of her student. "Trina, let's read the first paragraph," with which Trina jumps up and goes to the nearest door jam and starts beating her forehead against the wood. "Lisa did it! Lisa did it! I didn't tear the book. Lisa did it!" And begins to cry.

We also spent considerable time "teaching" the kids to set the table, fold their laundered clothes and bathe themselves, with approximately equal success.

One of the most alarming parts of the job was the apparent disconnect between the case workers who visited each week and their clients' conditions. It was as if the professionals had bought into the description we were given at the outset—that these children were only mildly disabled instead of being profoundly handicapped. The case workers controlled expenditure of Supplemental Security Income funds and deemed that they were always looking out for the children's best interests.

Case worker (Theresa): "I've noticed that Donny needs new shoes, his are looking a little scuffed. Would you take him to get a new pair?"

Me: "Sure."

Theresa: "And please don't get him cheap shoes, like Thom McCanns, he deserves a good pair. Here's a purchase order for up to $100. Don't spend less than $85." (That's the equivalent of $213 in 2009 according to the U.S. Government's Consumer Price Index.)

Another conversation:

Theresa: "Donny needs a new dresser. His dresser is shabby. Here's a purchase order for up to $200."

Another conversation:

Theresa: "Donny should be riding a bike at his age. Here's a purchase order for up to $150. I'm sure you can find a good quality bike for that amount."

And so "riding a bicycle" became part of the daily routine. I'd lead Donny outside and get him to grip the handlebars with the hand he wasn't chewing on. "Here's your new bike, Donny."

He would shove it away violently so it crashed to the ground. I'd pick it up, put his hand on the handlebars and say, "Here's your new bike, Donny." He would shove it away violently, etc. We'd do that for a half hour every afternoon and on the rare occasion when he'd actually grip the thing for a few minutes, I'd sign and say, "Yes," and "Good," before he'd fling it away again.

At least the SSI funds were supporting local businesses, I suppose. But the whole notion that we were providing a normal life was a charade. It was terribly sad and entirely hopeless, but it was what it was. Other than Trina, I don't believe that one of those clients had much cognizance that we were providing them with some version of family life.

Six months later we were judged to be very effective and reliable house parents and were recruited to take on a new home a couple hundred miles north in Chino Valley. We were thrilled, having made about as much progress as we were likely to make with our first crew, and—more to the point—we had camped and hiked all across southern Arizona on weekends from January through June and we were ready to spend six months exploring the red rock country, the Grand Canyon and other high points of the northern half of the state. An added bonus was the chance to audit some classes at Prescott College.

The Chino house was brand new, having been constructed by a wealthy doctor for the benefit of two of his children who were among our seven charges. The older daughter, Shelley, was in her early twenties and was always cheerful while the younger was shy and quiet and wheel-chair-bound.

At some point Shelley had been taught to make a corkscrew motion with a finger pointed at her head while saying "Crazy." She'd also make a drinking motion with her hand in the form of holding a cup while saying "Be-ah." It took us some weeks to understand that she was saying "Beer," at which time we obtained permission from her parents to provide her with a beer on Sunday afternoons, which she would sip with great delight while signing "Happy! Happy!" This suggested to me that she must have had experience of home life, or at least home visits during her earlier years.

Another memorable client, whom I'll refer to as Donna, was 19 years old with language skills at about a three year old level. She was angry and violent. Left alone with Lynn she would invariably dump her out of her wheel chair and laugh at the younger girl's inability to regain her seat. She repeatedly tore the heads off of Lynn's dolls and once grabbed me from behind as I drove the crew to school in nearby Prescott, in our household passenger van. Donna was very strong and was choking me as I wrestled the vehicle to the apron. She was given to disrobing in the boys' restrooms at the high school where she was being "mainstreamed," and by multiple accounts she was very sexually active in that situation, attracting a willing crowd of young partners.

So we "progressed," to the extent that progress was possible, going through the motions of teaching language and life skills to a household full of young adults with no prayer of learning language or life skills.

The crowning irony of that period involved Donna. Thanks to her sexual activity her parents had authorized a birth control pill prescription and she'd been taking them for four years (when she didn't spit them out). But she reacted violently to medical examination and didn't ever experience menstruation during the months that we were her caretakers. Based on the advice of her doctor who suspected tumors, and with her parents' consent, we scheduled Donna for a hysterectomy.

Wrong. Her case worker intervened, insisting that Donna was an adult, had a right to parent and that we were interfering with that basic human right.

Oh, indeed. She would be a great parent. Maybe tear the head off her baby? But we were legally outgunned at the requisite hearing. Donna was granted continued fertility. In the decades since I have occasionally pondered how the parents of young men in that high school would have felt about their sons lining up to have at it with Donna, perhaps gracing them with a grandchild, or, for that matter, how those now middle-aged men might feel about their first sexual encounters with a too-compliant mentally deficient woman.

In those pre-AIDS days they probably didn't pass around anything terminal, but there's little doubt that "safe sex" didn't figure in her picture. I still wonder from time to time whether Donna ever enjoyed the great satisfaction of parenting and whether any resultant baby survived. I fully understand the legal and ethical arguments concerning forced sterilization and the idea that we are all endowed with inalienable rights, but it seems pretty clear that there have to be rational exceptions to any rule.

My take-home lesson from that house-parenting year was that pre-natal diagnosis of severe mental issues and the availability of abortion-on-demand are one incontrovertible blessing of modern medical science and politics. None of the clients I tended that year had any remote prospect of living even semi-independent lives. Only the doctor's children enjoyed any measure of parental support (in that he built and partly financed the house) and even for those there was no meaningful family interaction. Families had abandoned the rest to state care, which, whether in large facilities or group "homes," amounted to warehousing. To the extent that giving over such fundamentally damaged children to state care was a painful choice, the families certainly suffered. But what enduring good was served by the entire effort? And how could anyone consider abandonment of a child to state caretakers to be more ethical than abortion?

Then, suddenly, we were free, our yearlong contract fulfilled, and we decamped to Tucson before a celebratory tour of Mexico.

During our stay in Chino Susan had shuffled vehicles. We bought a Ford Falcon van then swapped it for a VW squareback named "Buckwheat" by it's previous owner. Buckwheat had the same kind of pancake engine used in some models of Porsche and was fast as hell. One night, cruising home from a concert in Prescott, on the long straight-away that traversed Chino Valley, Susan took Buckwheat up past 90 mph. I was truly scared and asked her to back off, but she accelerated. I asked "What if someone pulls out, right now!"

She flinched and we started sliding in a high speed spin that took us 540 degrees, that is, once and a half around. We ended well off the road, in the desert, the rear hatch blown open and our college books and papers strewn along the roadside. Three tires had flattened, pulled off the wheels by the sideways friction.

"Did you see that?" she demanded, grinning. "I didn't roll her. I am fucking good."

A motorcyclist stopped to offer help, Susan told me to pick up the papers and change a tire, then jumped on the back of the bike and roared off toward home. By the time she came back in The Mother Van I had changed one tire for the spare and removed a second wheel. The next day we had them remounted and went back to pick up the car.

We soon purchased and outfitted a 1957 Olson Kurbside step-van and sold The Mother Van. The step-van was luxurious compared to the departed Mother, with head room enough for me to stand up when dressing. In memory of the departed Quinta (a.k.a. Big Mama) and because the Kurbside's aluminum shell had been painted in a shade of metallic green, we christened the new traveling vehicle "Quintaverde." So it was in Quintaverde with Buckwheat in tow that we drove south from Chino Valley.

Our inquiry into travel in Mexico had resulted in a few serious warnings: Purchase a dedicated auto insurance policy, but, in case of accident, attempt to settle with the other party before the police arrive—it will be cheaper for all concerned. If involved in an accident, be prepared to abandon your vehicle. Quickly. Pets are liable to be stolen and killed for food. Don't drink unbottled water.

So we procured insurance for Buckwheat and arranged for Susan's brother and sister-in-law to keep the cats during our journey. The afternoon before departure we had a going away party and following ingestion of a few margaritas I was convinced to shear my near-shoulder length hair and shave my head. Susan's brother, my age and prematurely balding, had been shaving his head for a few years and more or less dared me to do the same.

What the heck? It would grow back fast enough.

Hence I was just a day out of the shaggy haircut I had worn for more than a decade when I stepped out on the shore at Mazatlan and struck up a conversation with an older man in madras shorts and golf shirt that rode up over a hairy beer-gut. He turned out to be a retired police officer from Los Angeles and his conversation quickly turned to fond reminiscence about beating peace marchers bloody back in his heyday, during the war in Vietnam. He was genuinely enthused when he described the sound his baton had made when he made solid contact with a skull. "Just like kicking a goddam ripe watermelon," he said.

I knew in that instant that he would never have shared his story with me had we met two days earlier, and I resolved to continue shaving my head instead of letting it re-grow as had been the plan. The disguise worked pretty well in terms of fooling the opposition and I continued the tonsure for several years thereafter.

At first I underestimated the sensitivity of my newly bared scalp and managed to acquire a searing sunburn. In the absence of a useful quantity of clean water I opted to shave dry and one of the more memorable experiences in my life was dry-shaving a sunburned pate while being chewed alive by no-see-ums on the seashore in San Blas.

Decades later I read the story of five fishermen from San Blas who had run out of fuel and drifted across the Pacific for more than nine months before their rescue by a Taiwanese tuna boat in August, 2006. The writer made mention of the home port and the fact that unlike other seaside towns on Mexico's Pacific coast, San Blas had never become a tourist destination due to the hordes of biting gnats. True that. I have never been so viciously chewed, neither by swarms of Everglades-, nor hordes of Alaskan-mosquitoes, nor in black fly season in New England nor in wet Minnesota woods nor Louisiana bayou bugdom. Not even upon return to a closed-up house wherein a million-billion fresh-hatched fleas lurked in their patient hunger. Sanity seems a remote possibility amongst such tormentors.

Aboard the drifting vessel the three survivors ate whatever sea life they could get their hands on, literally jumping overboard to ride and wrestle sea turtles. They collected rain water in empty gasoline cans and read the Bible aloud while they drifted and prayed. Their rescue reportedly confirmed their belief in prayer, given as most humans are to resort to superstition when in dire straits. There was no report on the efficacy of faith for the two companions who died and were dumped overboard.

The bigger lesson might have been to plan ahead next time.

Mexican gasoline was problematic for auto drivers too, more due to poor quality storage tanks than to failure in refinery. Although we had been warned to only purchase premium grade fuel, we weren't told to change fuel filters on a regular basis.

We were driving on a two lane highway through a mountain range some hundreds of miles further south when Buckwheat coughed and died. We had been clenching our teeth and swearing as we met oncoming chicken buses and freight trucks whose drivers seemed blithely unaware of our presence, and now we were stalled on the narrow roadway with no shoulder, a sheer cliff falling away on the other side of the pavement and darkness fast approaching.

We had just long enough to carefully analyze our situation ("We're screwed.") when a bright orange VW Beetle stopped about fifty yards to our rear and four large men in dark glasses piled out and started advancing toward us. ("Oh, great. Now we're going to be robbed.")

But no, as it happened we were about to be rescued.

"Buenas tardes, my American friends," said the incredibly handsome driver, who sported a gold chain and an expensive looking watch. "¿Que pasa? What seems to be the trouble?"

"It just quit," I answered.

"No problema!" he responded. "We push."

The foursome returned to their vehicle and drove up behind Buckwheat. The bumpers matched exactly and we jumped back in.

Our unnamed savior gently nudged us up the mountain, through curve after curve until we arrived a small farm house with children dashing about and chickens scattering at our approach. Two men got out of the car and another jumped in and we were off again. Soon the road straightened as we exited the mountains and drew closer to Puerto Vallarta. We were pushed at speeds sometimes exceeding 60 mph for more than 45 minutes, during which the other driver stopped to unload and load passengers then caught up, eased into contact and continued on our way.

Finally he spotted one of the tourist service trucks known as "Green Angels" and flagged it down to direct it to our aid before waving goodbye and speeding away. The helpful mechanic changed out our fuel filter in just a few minutes and we were soon on the road again having been admonished to purchase a few spare filters at our earliest convenience.

Two days later while walking in the city we were hailed by our helper, who flashed a million dollar smile and called out, "Buenos dias, my American friends! Have a good journey!" He had swapped the orange VW beetle for a yellow Cadillac convertible and he had swapped his male passengers for three stunningly beautiful women in bikinis. He looked like a movie star.

During our stay in Puerto Vallarta we camped in the parking lot at the ferry landing, having been told by other budget-minded tourists that we were unlikely to be hassled since the ferries to Baja California were infrequent and hanging out there was common. "Just say, 'Esperar el transbordador.' and you'll be okay."

We had been in Mexico for three weeks and had spent a good bit less money than anticipated, so we decided to treat ourselves with a dinner out instead of our usual camp food. The tiny restaurant we selected had just three tables and we were the only customers of the woman who was both cook and waitress. We asked what was best on a surprisingly long menu and took her advice.

She served us each a Carta Blanca, then doffed her apron, picked up a shopping basket and dashed up the street to purchase the requisite ingredients. The food was superb and afterward we further splurged by purchasing a bottle of mescal and polishing it off while sitting in beach chairs at a posh ocean front hotel. Susan, ever the daredevil, even ate the worm.

We stumbled back to Buckwheat sometime after midnight.

Susan was unrelievedly compulsive about accounting for expenses and jotted down every expenditure over the twenty-five years we coexisted. That was how we knew to the penny what we had spent to that very moment in Mexico (in both pesos and dollars) and she insisted that we write down the evening's expenses before retiring for the night.

But we couldn't find the ledger book. It wasn't in its usual location in the passenger side pocket. It wasn't under either front seat. Nor behind the seat.

Now, understand that we had each imbibed a generous portion of mescal and were happily lit, and just tipsy enough that the search turned into kind of a fun game. We pulled out backpacks and the sleeping pad and cartons of food and then had the hood open (the storage compartment in a rear-engine VW) when a hand fell on my shoulder.

I turned to face two Federales, or perhaps port guards, dressed all in black and cradling large automatic-looking weapons. Kalashnikovs?

One of them fired a string of Spanish at me that quickly surpassed my half-remembered high school language skills.

Envisioning months chained in a subterranean jail cell, eating rats and begging for a chance to contact the U.S. embassy, I groped for an explanation. Bam, I hit on a word from my very first Spanish lesson, way back in eighth grade. "Cuaderno." Notebook.

I picked up my journal from atop Buckwheat.

"Este es un cuaderno," I said, offering a friendly grin and doubtless slurring the old Español. "¿Si?"

The two officers looked warily at each other and nodded.

"Tengo dos cuadernos, pero solamente veo uno," I said, feeling more confident as I explained that I had two but was only seeing one. "¿Donde esta el otro?"

They looked to each other again and shrugged. Like they were supposed to know where my other notebook had gotten off to?

I remembered the excuse we'd planned concerning the ferry. "Esperar el transbordador," I added, hopefully.

Bingo. My get out of jail free card worked.

"Ah. Buenas noches," they muttered and walked away shaking their heads at the idiot turistas.

In the ensuing minutes of relief and clarity (what is it they say about a death sentence concentrating the mind wonderfully?) I remembered that immediately before we headed out for our night on the town I had tossed a bag of trash in a nearby garbage can. I retrieved the bag and there was Susan's ledger.

I guess she was pretty relieved to get it back because she didn't punch my arm as hard as usual when she called me an asshole for losing it and for nearly getting us arrested. Then she wrote down the night's expenses while I reassembled our gear in the car.

Meanwhile, just offshore, grey whales were basking and blowing, together with their newborn babies.

light

If the sun refuse to shine. I don't mind.

Nah. Jimi Hendrix was wrong. The approach of the winter solstice is a time that "tries men's souls." Women, too, find their hope waning with the light. We find ourselves unabashedly thirsty for wassail or a manicure or a new DieHard or an iPhone to make things right. Even the mercury begins to look a little low. Here, on top of Big Blue, the days get short: darkness leaks into mornings and afternoons. In the trough of winter, twilight becomes the norm.

There is a tiny although presumably measurable statistical possibility that the trend will continue. As the sun dwindles, it may get darker and colder forever. The seasons seem to follow each other in a circle dance, but, hey, you never know. Stuff happens. There's the physics we know and the physics we don't know, and as of this writing we still haven't figured why time only goes one way. Einstein showed us that it ought to be reversible, and scientists now speculate the answer may come from dark matter.

Dark matter. Of course.

In the weeks leading to December 21, possibilities fade. Life itself seems problematic. It is no surprise that religions ancient and modern attach such great importance to this time of year, when Christ, Hindus' Krishna and Mithra (the Greek God of light) share a birthday. Eat, drink and be merrily pious for tomorrow ...?

"The mercury," I said, "is low," referring to that queerly self-absorbed, cohesive and toxic metal of thermometric fame. But consider Mercury, the planet.

If the human race had arisen on the first rock from the sun instead of the third, we would observe no solstices. Mercury's orbital axis is vertical with respect to its orbital plane. Every day is the same. Every season is summer. Would there be religion?

Okay, okay, I know it's hot as blazes there, and Life-As-We-Know-It won't float. (No water, for starters.) But, take my question as a point of departure. What if Big Blue's orbital axis were vertical too? Without a season of despair, would there arise an organized effort at supplication and hope? If things didn't look occasionally bleak, would a savior be of any particular interest? Without seasonal extremes would the very notion of heaven and hell have meaning?

I don't think so. If life were like elevator music—never better, never worse—who could postulate choirs of angels?

And so the Winter Solstice, freighted with Zoroastrian astrological baggage, Wiccan dance, Druidic chant, Hebrew salvation, Christian, Greek and Hindu birth, and New Age reconfiguration: children of the northern climes, all. South of the Equator, our shortest day is the longest, our shivering blizzards a far cry from their lazy, hazy, crazy days of summer.

No surprise, then, that the northern faiths have proved something of a tough sell below the tropics. Not only does our hardest hitting holiday fall in the wrong seasonal slot, but vast oceans so temper the clime that there is no big incentive to seek salvation. Down under, native beliefs tend toward animism and pantheism, dreamtime and songlines. (It is probably easier to feel at one with nature when you can comfortably run around naked most of the time.)

None of which diminishes the importance of the season for us. If it takes a little darkness and cold to remind us to be thankful, let's start with thanking nature for putting a little tilt in our wobble. If trees which found their ecological niche by retaining leaves through the dark season became a pagan symbol of life, let's bring an evergreen inside and honor the miracle. If gift giving sprang up as a way to share summer's bounty and insure survival of the clan through snowy days ahead, let's find that true spirit of giving—selfless charity. If solar re-birth heralds maximum seasonal potential, let us look to a baby for inspiration and hope.

We have, all of Big Blue's children, found our separate explanations, placed our trust in separate gods or the verities of science, and striven mightily to convert the heathen or disabuse the believer. We have hidden self-serving motives under trappings of religiosity and fought crusades heralding truth while we pillaged.

When we edge past the solar nadir we can sit before the fire (a bit of photosynthetically trapped sun-shine) and light a candle (more solar power: beeswax or whale oil or dinosaur juice) for peace. We are sisters and brothers on this big blue ball. As the mercury dips, let's huddle for warmth, and remember what we share.

I began this illuminated detour in song and end with another. Joni Mitchell this time; lyrics for the solstice.

> There'll be new dreams, maybe better dreams,
> There'll be plenty of new dreams, before that last
> revolving year is through.
> And the seasons, they go round and round ...
> And round in the Circle Game.

10. sarajevo

I was in a low-down funk.

We were high up in the mountains of western Virginia, and camped in a place called Raven Cliffs. The plan had been to return to New Hampshire for a few months but Quintaverde's clutch went south as we ascended the Interstate north of Statesville, so we'd detoured into Mt. Rogers National Recreation Area to consider our options.

But the clutch wasn't causing my funk. The dark mood was more existential. I was 37 and not particularly thrilled with the direction my life was headed. Susan's insatiable travel bug kept us moving and moving and moving but we didn't seem to be getting anywhere. Not that I could have told you this at the time. Denial is a deep river.

By this point we had the cabin in New England and a small house in Black Mountain, North Carolina—which had come to feel like the only real home I'd had in my adulthood. More and more I wanted to be settled, to start doing whatever it was I was going to do with my life. In the five years that we'd been based in Black Mountain I had begun to grow roots, a strong sense that I would live out my span in the Southern Appalachians and ...

And, what?

The building trades are immensely satisfying in a very basic, tangible way. You make something where there wasn't something before, or you set things right where they have sagged or rotted or settled. It's a constant study in applied mathematics and thoughtful finessing of approximations to meet design ideals. Home Repairs merit badge and the Pythagorean theorem were my stock in trade. I was and remain proud of my work and I will always have enormous respect for good builders. So I wasn't exactly unhappy with my job, but, in retrospect I understand that there was something missing.

The gardening that occupied so much of my life was and is viscerally satisfying as well. There is nothing in my experience that forms as tangible a connection to the cycles of life and deep ecology as putting ones' hands in the dirt. While growing plants you also grow in under-standing of nutrient cycles, soil life, the habits of pests and plants, worms and weeds, rodents and serpents, owls and hawks and chickens and rabbits and much more. But, again, there was something amiss.

I was well into my middle years and felt more than a little directionless, though I would surely have told you I was happy. When I'd dig just a little into my sense of ennui, and try to explain the feeling to Susan I was met with disdain. "Oh, so you're not famous yet, is that it?"

"No. It's ... I don't know. It's probably nothing." Now, as then, I'm often less than clear about the emo-tions that move beneath my surface, though perhaps I have learned to consider that such unseen, unsensed movements are going on somewhere inside. I know I like to think I've learned a little along the way. Don't we all?

Sitting at the picnic table there in the Raven Cliffs campsite, Susan was fed up with my immediate moping. "What the hell do you want to do?" she demanded. "Just tell me what the hell you want to do!"

"Whittle," I replied on a sudden whim. "Whittle."

Susan jumped up, walked to our stack of fire wood, grabbed a short limb and slammed it on the table in front of me. "Then whittle, damn it! I'm going for a hike."

Thus began a few years in which I was never far from knives and chisels as I carved a series of small human figures alternating with wooden spoons. That afternoon I made substantial progress on the figurine that would show up in verse five years later when I was slamming poetry.

Little Whittled Wooden Women

I am enamored of wood and when left alone am
prone to whittle little wooden women.
When my best friend saw the first of these
(a hiker with a knapsack) she said,
"Hey, you've got a knack, sap!"
I told her it's a snap, that
I can see the figures in the grain
and carve to set them free.
A tree remembers.
That old trunk holds memories of lives long past
like childhood photographs.

It is clear tonight.
A chill wind rattles bare branches as
I toss another stick of furniture in the fire
and read Frost,
Doing neither lightly.
Respecter of wood and figure
and limb and craft,
I would not feed the stove this chair
Unless weathered and weakened beyond repair.

Fire and ice, posit the poet,
Stand opposed.
Potential conclusions to our little adventure.
But reading Robert by my flaring chair
My mind is drawn to Sarajevo
Searching for the figures there.

This post-Cold-War winter has proved
post neither in Sarajevo. Ice has demanded fire

and the trees of the city have been cut and
burned.
Ice has demanded fire and the furniture in the
homes has been given to the flames.
Ice has demanded fire and the books on the
shelves have been heaped into the blaze.
And, one must suppose, the little whittled wooden
women as well.

What surreal triage that calls to mind!
Hey honey, what do we cook on tonight?
Your piano or my guitar? Value versus
portability, no small issue in time of war.
Do we burn the dictionary first?
A lifeless list ungraced by passion, or last?
That oh, so essential key to whatever library
remains.
And if we choose to warm the baby's food on
Noah Webster's fire, do we start with "A" or "Z"
or Foreign Words and Phrases?

When a city burns its trees it gives up shade
and beauty and the juicy fruit of summers
yet to come.
And with the furniture, comfort and utility are
sacrificed for survival.
Then with the books, the hard-won hope of
history is given to the flames and
With the little whittled wooden women,
what loss? Memory, perhaps? Of a time when
there was time to look for figures in the grain?

I set poetry aside. Pick up knife and chunk of
cherry.
There's a little wooden woman there who needs
to be set free. To live awhile on my shelf.
To lend her warm peace to this room.
Waiting, waiting, always waiting,
For the day when ice cries out for fire
And memory is a luxury we can no longer afford.

All but one of those figures are elsewhere today, given as gifts, perhaps later tossed, or, who knows?, burned? The remaining little whittled wooden woman depicts Susan wearing a bandana on her head, in her toolbelt, hammer in hand and with her trusted familiar, Snook, draped across her shoulders. Though it is not the particular figure I described in the poem I wrote on one brisk winter evening while actually burning a decrepit chair in the wood stove, it is carved in cherry and perhaps lends a warm peace to my home, a warmth and peacefulness that never emanated from the model. A missing warmth and peace I couldn't have explained to myself if I'd tried, and a warmth and peacefulness I must today admit I never engendered myself.

However, as the present work attests, I have found that I can still afford the luxury of memory.

I suspect no place will again represent "home" in quite the personal way that the house in Black Mountain grew to fill that totemic niche in my soul. The simple fact of living in one place for two decades, or at least being based there through those peripatetic years, is most unlikely to repeat itself, given my age as I write this.

But the connection was much deeper than the tock-tick of my biological clock and the race-track pace of the calendar. I felled oak trees, split them with axe and maul and wedges and hewed them with an adze* to make the carrying beams for that structure, then hauled them with block and tackle to rest on foundation piles I had laid up and poured solid with concrete and steel. I fashioned a froe* from an old automobile leaf spring and

***

* An adze is a kind of sideways axe, similar to, but much sharper than, a mattock. A froe consists of a blade about sixteen inches long, sharpened on one edge along the entire length and controlled with a wooden handle which forms an "L" shape. Dogwood is among the hardest native species and the root the toughest part of the tree. A dogwood mallet is used to drive the froe, edgewise, into a length of wood to split off shakes.

hewed a club from a dogwood root to split shakes from chestnut oak blocks for siding. I salvaged wood from buildings which were being razed in three surrounding counties to raise my walls, and salvaged cedar planks from a demolished deck to add one to my house.

The water system I created consisted of pipes from the gutters on the second floor which ran into the walls (to prevent freezing), into the ground and then uphill to a 1,200 gallon concrete tank (sold as a septic tank) buried in the hillside with the bottom about level with the ceiling on the first floor. Because the top of the tank was below the height of the gutters, the water flowed through the "U" shaped pipe and into the cistern due to gravity.

Then, because the bottom of the tank was above the first floor plumbing, the stored water ran back down to the sinks and bath tub—thanks again to gravity. This system was a principle key to our living off the grid for two decades. Water pumping is often the trigger that tips those who seek energy independence into the arms of the electric utilities.

The sinks and tub were second-hand as was the "pissoir," an indoor urinal that was the companion to our composting outhouse. Most composters smell bad because of an excess of urine, so the best way to avoid the stink is to pee elsewhere. I created the pissoir for the benefit of women: whereas guys can simply piss in the woods, anatomy makes that more problematic for gals, exposing more of themselves to the elements and onlookers. It consisted of a stainless steel sink salvaged from a Greyhound bus, which was exactly the right dimensions for a standard toilet seat. A small tube and a valve provided a flush mechanism and the outlet went through a standard trap into the grey water drain field (along with tub and sink waste). Used toilet paper went into a brown bag that was burned at necessary intervals in the wood stove.

In the twenty years that we relied on rainwater for household use we only ran dry once, and that was due to a broken pipe, frozen during our absence in one very cold December. (Drinking water was hauled from a spring, not on our property.)

I wired the house with standard outlets, switches and circuits, although we ran the place on 12 volt DC power until 2000 (from a battery bank charged with photovoltaic panels). When Susan's uncontrollable shaking and shivers (due to chemotherapy) required electric blankets the only practical option was to hook up to the grid. Purchasing enough solar panels and batteries to handle a high demand load like electric heating is, of course, not an impossibility, but by then Susan was talking about selling the house and I was in no mood to contradict her wishes. At about that time, a friend's difficulty selling an off-grid solar powered house in the next valley suggested that a grid hook-up would find a wider market if I did accede to the sale. Because of my initial standard wiring system the switch to grid power was as simple as disconnecting the batteries and running a cable from the new electric meter to the circuit breaker panel.

Many of the windows were recycled and most of the sheetrock was salvaged cut-ends from job sites. Most of the light fixtures were home-brewed, made from salvaged sockets and switches. Even the bathroom and kitchen tile were recovered from a tear down.

Aimed 15 degrees west of true south to take maximum advantage of solar heat, with no windows on the north walls and substantial glass on the south, much of the winter heating was via passive solar. A wood stove provided back-up, and a propane wall furnace located near the plumbing core provided back-up-back-up for the occasional winter absences when we didn't shut down the water and drain the pipes.

The house was low-tech, low cost, very green and very personal.

In using the first person pronoun throughout this description it is not my intent to diminish the work Susan contributed to that house. She did some of the physical chores and made many of the interior design decisions. But the fact of the matter is that I performed most of the work, often when she was away, and created the mechanical systems more or less from scratch—which made the building very much my project.

Furthermore, it was never home to Susan, who lived her entire life referring to Canton, Ohio, as home, even to the point of argumentativeness. I might suggest that I'd prefer to be home for Christmas and she would enthusiastically agree that we should plan to be in Canton. Or, during a stay in New Hampshire I would say that I wanted to head home soon and she'd insist that , no, she wanted to head back to Black Mountain. When I'd say, happily, that I had never felt so much at home as in that place, she'd tell me it was a piece of junk and that no one could ever regard it as anything more.

That fixation on her childhood home versus our shared home was unquestionably deep-rooted. Susan's family was close-knit, with two aunts and an uncle settled within one suburban block (the former family farm, later sold off for development). Two of her three brothers lived within several miles, and other members of her extended family dotted the landscape. Moreover, her father had succumbed to a heart-attack when she was just sixteen, a loss more keenly felt because that was the parent she most adored. His death rocked the family world, cancelling her fully formed college plans and sending her into the working world the following year, while her relationship with her mother soured.

That paternal absence was a black hole in her life with a gravitational attraction that constantly drew her thoughts. No month went by over the decades I knew her that lacked some specific reference to her "Daddy" and his untimely death.

Apples among her clan landed close to the tree. In my much smaller family, ties had been unbound for a generation. When I blew away from home, like dandelion down on a strong breeze, taking root seemed the natural thing to do. I tried it first in Florida where I staked out a homestead in a palm and live-oak hammock, later in New Hampshire in Mountainview Swamp , and once again in the Southern Appalachians where I've spent the last three decades, I was eager to work my feet into the dirt and get a grip.

All the talk of my hand-made home's junkiness didn't seem to recur to Susan when she transferred title to the house to her nephew and had her attorney order me out in 2001. Of all the many things she stole from me in our finale, taking my home took the cake.

But that was far ahead as I sat at Raven Cliffs quite contentedly whittling a little backpacker, her mouth puckered in a whistle, just as mine so often was when I hiked alone. In a few days we headed back to North Carolina, following the route with the fewest steep uphill runs to baby the clutch until we were back on flattish ground with a machine shop in easy reach.

fire

Your life reruns when you're dying.

Or, at any rate, mine did when I thought my bucket was about to be permanently and irrevocably kicked.

The first go-round I was 8 years old, so the rewind was pretty brief. Removal of tonsils was in vogue at the time and while I lay on the operating table, cruising along in my ether-induced trance, a red light seemed to pulse above me. (The glare of operating room lights through my eyelids was fluctuating due to the rhythm of blood in my veins, I suppose.) With each pulse a litany within: "You're dying, you're dying, you're dying, you're dying."

My thoughts raced through a brief montage of experiences and I felt terribly, terribly sad that I would never see my Mom again. Soon enough I woke up in a recovery room with a mildly sore throat and ate jello and ice cream for a few days.

However, that scene replayed just as I drifted off to sleep for many months after. "You're dying. You're dying." I'd jerk awake, and stare at the ceiling, wracked by fear for a few moments until I got my bearings.

The second time had a similar residual effect but it happened so fast I barely had time to blink let alone go retrograde.

It was 1968, I was 17 and my girlfriend and I had attended the Winter Park High School Junior/Senior Prom, joined friends at her parents home for a midnight breakfast and then headed to the beach in my Dad's brand new 1968 Ford Custom—an unusual treat accorded me for the special event. That V-8 engine ran like the proverbial bat out of hell.

We were good kids. No alcohol or drugs or sex. We parked on the beach at New Smyrna near the north inlet, made out for a while, which in our good kid way involved kissing and hugging, and then laughed our way to dawn. We swam and played with a Frisbee and talked with our crowd of friends until mid-afternoon, then headed home —about an hour away.

The all-nighter and the day's activity and sunshine were catching up with us and my girlfriend yawned and curled up on the seat beside me and fell asleep nestled under my arm. I woke up doing 60 mph headed dead-on for the concrete base of a light pole and instinctively veered hard right. The rear bumper just clipped the pole leaving a pale half-inch scratch in the chrome, small enough that I never had any explaining to do to my Dad. I, on the other hand, was pierced to the core, trembling, in a cold sweat. My girlfriend, barely jostled, half awoke, asked what was up, and settled again when I assured her than everything was okay.

But for a couple of years afterward, more nights than not, just as I drifted off to sleep I'd suddenly face that light pole and jerk awake, completely terrified.

The most recent go-round with mortality happened forty years later after I'd heard my fair share of near death accounts from others, the most haunting of which came from the recovered cockpit voice recorder of a doomed commercial aircraft. A choked out, "Mom, I love you ..."

End of tape.

The oddest account came in 1982, about a year after I moved to North Carolina. I was sitting in our Volkswagen Squareback with a canoe on the roof, in a grocery store parking lot, when a grizzled old mountain man approached and gestured at the canoe.

"Ever been to the Nantahaley Gorge?" he queried, in a reedy voice. (The Nantahala Gorge is famous for whitewater rafting and kayaking, but at the time I'd not heard the name.) When I shook my head, he added, "You should. You should." Pause. "I died there, you know."

"Oh, yes?"

"I died three times; three times! But each time the Lord brought me back. By water and by lightnin'. It was the Powers that saved me." He needed no prodding to continue. "The first time I drownded. We was building the dam up to Fontanney [The Fontana Dam was completed in November, 1944] and I fell in the river.

"I can't swim, you know.

"I sunk straight to the bottom, plumb like a rock, and I drownded. I could see the people on the other shore and they were wavin' at me, tellin' me to come on and join 'em, and I saw my Daddy and Mamaw and all the others, and it was so peaceful and nice. But a voice said, 'It isn't his time,' and that was the Powers that was speakin'. And two fellas pulled me out of the river and laid me out on the shore."

"Now the second time was the lightnin'. I was lyin' in my bed and the lightnin' came in through the window and took me and kilt me dead. But then there was a window in the other wall where there wasn't a window, and I could see all of them over on the other shore wavin' at me callin' to come on over with 'em. I could see how peaceful they were, but an angel come in the window where there wasn't a window and said, 'It isn't your time," and I knew it was the Powers and then I fell asleep."

He subsided, so I asked, "And the third time?"

He scratched his head, "The third time was when we was buildin' the Fontanney Dam, what I already told you about. It was the Powers that saved me." He looked at me as if he thought I hadn't been paying attention and shook his head. "You know, if I had that canoe, I'd go on over to Nantahaley and I wouldn't worry about fallin' in, on account of the Powers has told me that I won't die until it's my time. Nice to meet you sir." And he walked away.

Seventeen years after that I was the volunteer keeper of roads in my small mountain community. My chores involved figuring out where we needed machine work or gravel to keep three miles of unpaved road passable. When erosion demanded remediation I'd call in a few truck loads of "road bond" (a type of crushed rock that includes "fines" which help make the surface smoother and more compact).

One afternoon as my delivery driver started to spread 15 tons of gravel, his third load of the day, I sat beside him in the truck cab and started to write a check. He raised the dump bed to full height, released the tailgate and started forward. The right rear wheels rolled up on a massive boulder that protruded a few inches above the road bed, the truck swayed, lurched, and suddenly we were tumbling to the left, down the mountainside. The interior of the cab looked like an astronaut training session inside a diving plane. Lunch box, hard hat, tape measure, pens, pencils, jackets, check book, me, the driver, floating and banging around as we tumbled.

During those long moments my principal coherent thought was, "Oh, so this is how I die. Now I know."

I experienced a montage of scenes from my life, good and bad, seeming to remember everything at once, but with some sense of sequencing them, as if I were setting things in order. I remembered the tonsillectomy and the Powers, biting gnats in Mexico and butchering crabs in Alaska. Songs I'd written and the poetry I'd slammed. Radio shows and saw-toothed mountain ranges. Christmases and birthdays and my wedding and my favorite cat. I felt terribly sad that there was no chance to say goodbye to the people who had mattered so much. "I love you," I thought. "I loved you all so much." Still, there was a great sense of calm and confidence that life would go on without me.

Sure, there were some regrets for things I hadn't gotten around to doing, but a profound understanding that since I was about to disappear from the universe, there would be no me to regret anything, and that was just alright.

Metallic grinding and the sound of tree trunks snapping overtook the engine noise as the multi-ton dump truck rolled completely over and half way again before lodging against stouter trees, upside down, with the driver and me tangled and settled on the inside roof. The truck swayed as the trees that had stopped us bent under its weight. I shoved my door open, the truck shifted. The driver groaned.

"Okay?" I asked, he nodded. "Here I go," I said, and swung out, gripping the edge of the roof and dropping ten feet to the hillside where I tumbled, righted myself and got out from under the truck as quickly as possible.

Moments later the driver followed, groaning loudly as he landed. Later we'd learn that he had three broken ribs. I helped him scramble up the hill, both of us bruised and battered, and we limped our way to the nearest occupied house with a phone.

The truck was recovered at great expense to the insurers and mechanics reasoned that the lift piston had snapped, causing the sudden load shift that took us down the mountainside.

Taken altogether it was a cheap lesson on dying—a full-on body bruise and a few moments of retrospective regret.

Some would say that my "near death" experiences don't count, because they aren't the same caliber as those whose hearts actually stop beating, dip into coma and return, lose a whole lot of blood or die on the operating table only to be resussed. Could be, but it sure felt like the end of the line where I was sitting, and death wouldn't scare me ever again.

Not that I wouldn't look for immortality. After all, I'm a writer and like most (all?) writers there is some element of that desire to endure in what I do for small pay and at sacrificial length. But corporeal death will come when it will come and I wouldn't want it to be any other way.

On the other hand, I'm not eager to die.

Consider this: If you knew you would only have fourteen thousand dollars to last you the rest of your life, chances are you would think twice about expenditure. You would know that a dollar spent was a dollar gone forever. You would probably buy seeds instead of food, cloth instead of clothes, a bicycle instead of a car; generally choosing the durable over the throw-away, and the easily maintained over the complex.

Replace dollars with days, however, and we are readily spendthrift.

Our little blue planet probably won't spin around more than twenty thousand times before you are outta here. If one includes children, Americans have an average life expectancy of just over fourteen thousand days. That's fourteen thousand more sunrises and -sets; a little more than five hundred full moons, and perhaps eighty solar eclipses. Only a handful of visible comets will swing by before you go, and about as many spectacular meteor showers.

At age 59 I'm on the downhill stretch. With reasonable luck daffodils might pop up twenty more times before I am pushing them up myself and autumn leaves will only swirl around my feet on something like six hundred more days before I am pulling them up around my chin to snuggle in for the winter.

The brievity of life is a commonplace, of course. There's nothing particularly profound in noticing its pace. Even that relentlessly material girl, Madonna, once told *Rolling Stone* magazine that "I feel the fleetingness of time. And I don't want to waste it on getting the perfect lip color."

It's easy to give lip-service to the years, but it's the days that slip by. A year is long enough to be something of an abstraction. We reach another birthday and say we don't feel a year older. We make a list of resolutions on January first, and know we'll have plenty of time to make some changes. But then we spend a day shopping and another getting the car repaired, a morning at the dentist and another day recovering.

My best days seem to pass by in a place completely out of time. A day to paint a still-life. A day to read or write. A day to hike into and out of unknown places.

When Shakespeare observed that "tomorrow creeps in this petty pace from day to day," he was not describing the modern experience. If Shakespeare's days crept, it is clear evidence that human consciousness has changed radically in four hundred years. Our days are better described in the Beatles' "Day in the life." "Get up, get out of bed, drag a comb across my head. Find my way downstairs and have a cup, looking up, I noticed I was late."

Some folks call Wednesday, "Hump day" and say it seems slow, but the weeks speed by. Deadlines race into view and nearly knock us over as they sweep past. No sooner are bills paid than another round pop out of the box.

And that is where things begin to blur. Months teeter on the edge of graspability. I could easily describe to you what I did last week, but last month demands some generalization. When we announce plans for tomorrow they tend to be definite. Plans for next month are more like good intentions, with perhaps a couple of contractual obligations and a lot of blank space between.

Recent research suggests that we have an innate time-line built into our brains' circuitry. Our basic understanding of numbers is stored as visual information; line segments of differing lengths. This concept stretches forward and back, giving us our sense of past, present and future. Vision involves perspective. A distant telephone pole is no bigger than your finger, and mountains on the horizon are shorter than your thumb.

But I don't want to let the apparent distance fool me, or the abstraction of years muddle my view. I have about 7,300 days in my account as I write this, and many fewer by the time you read these words.

For those who believe that our spirits survive death in some way, death may seem unimportant, even something to be desired. There's no doubt that survival and heavenly reward constitute one of the strong attractions of many forms of religious belief. But all of the evidence I've seen so far suggests that the end is the end. Those of us who believe that this life is the only one we will be accorded are bound to hold our lives to be more significant. I would even posit that most atheists consider life to be more "sacred" than do believers in spiritual survival.

It's often said that there are no atheists in fox holes, an idea I have no basis to contest. But I know for certain that if I were a Private in a fox hole I would "pray" that there were atheists in the company headquarters. If I thought I were likely to be sent to my "just reward" by a commander who believed in never-never land, I would be forced to abandon my post. It seems to me that such religious beliefs are what permit generals and presidents, congresses and parliaments, kings and queens, in the good of their own consciences, to treat privates as cannon-fodder.

In this regard it is disturbing to reflect on the report by French President Jacques Chirac that George W. Bush, told him: "Gog and Magog are at work in the Middle East ... The biblical prophecies are being fulfilled ... This confrontation is willed by God, who wants to use this conflict to erase his people's enemies before a New Age begins." So it seems that a politician's religious fervor rather than illusory WMDs was the rationale for ousting Saddam. Atheists might take their nations to war, but at least they don't delude themselves concerning divine guidance or assuage themselves with pleasant illusions about the outcome for the deceased.

Non-theists are underrepresented in prisons and divorce courts and countries with the highest percentage of atheists are statistically safest. About half of Europe's population are non-theist and Pew Research reports that countries with the highest levels of non-theism have the lowest homicide, poverty, infant mortality and illiteracy rates. Such statistics lead me to the supposition that we who eschew religion take life, commitment and perhaps ethics more seriously than our worshipful cousins. I well recall Bobby, the only Catholic boy in my grade school class, who taught the rest of us curse words and dirty jokes and assured us that he had no trouble with such naughty behavior because he went to confession every week and was forgiven. Which is not to suggest that grown up Catholicism is as naive as a child's, but he seems to have nailed the basic theory.

All told, I plan to make this day a good one, because I don't believe I'll ever get it back.

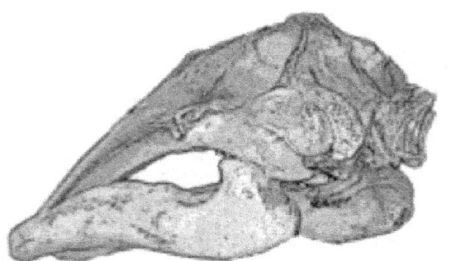

11. skulls

Brave Ulysses loved Mountainview Swamp.
When we returned there after months and even a year away he would make a beeline for the cat door and be preening on the porch long before we retrieved the key to the padlock that permitted human access. My favorite, albeit blurry, picture of Ulysses is in his preferred perch on a wooden toolbox on that tiny porch.

Beginnings and endings being as broadly arbitrary as they are, it is arguably appropriate that he was gunned down there in his woods, versus whatever other place he might have ended—a couple or a dozen years further on. His murder occurred the day before the "final" broadcast of "A Prairie Home Companion," in 1987. The radio show was later resurrected, and, of course, Ulysses has stayed quite dead. I remember crying through much of that broadcast, with Garrison Keilor singing:

One more time, this dance together
Just you and I, now don't be shy
This time, I know I'll hear the music
If you would hold me one more time

That yard, that tree - you climbed it once with me
We talked of cities we'd live in someday
I left, old friend, and now I'm back again
Please say you missed me since I went away

It may seem maudlin, but the sense of loss and an

end to good times was all too palpable with the passing of my dear, dear companion. More than twenty years on, I still tear up at the memory of that song and that amazing cat who knew my moods before they came clear and evinced the introspective demeanor of a sphinx.

Skulls endure as visible testimony to vertebrate existence, and empty eye sockets seem to me to be what most attracts or repels and sometimes even haunts us when we examine the empty shell of another's life. That's taken together, of course, with the stark foreshadowing bare skulls offer concerning our own mortality. It's the source of power summoned by a pirate flag and the horror summoned by scenes from Cambodian killing fields or the excavation of mass graves. It may even be the source of our ancient preference for burial or burning of the dead, to remove the stare of unseeing eyes and a reminder of the departed.

All through my life I've been fascinated by skulls in an amateur naturalistic sort of way. In grade school I dissected owl pellets, the regurgitated remains of digested prey which enable raptors to consume undigestible animal parts without damaging their intestinal tracts. Almost every one contained the skull of a mouse or bird, the ivory shell seeming somehow at odds with the shape of the original owner's head, absent flesh and fur or feathers. Playing in vacant lots I'd come upon larger skulls—of rabbits and raccoons and cats and dogs, or so I might guess based on size and dentition.

Later I'd use *Peterson's Field Guide to the Mammals of North America* to identify beaver and deer and opossum and, later still, a seal skull on a beach in Maine. I would guess that Ulysses wasn't buried, unless the gunman elected to hide the evidence, so his skull may be out there in the wet New Hampshire woods waiting to fascinate some future ten-year-old or slowly dissipate into constituent chemicals and be absorbed by a beech or yellow birch or the oaks and hemlocks that populate that small mountain.

In *The Night Country: Reflections of a bone-hunting man* (Charles Scribner's Sons, 1971) I learned that the anthropologist Loren Eiseley shared my fascination. He wrote, "I get out all the skulls. A massive unknown cranium which bears the look of the Cro-Magnon past about it is one I rescued from a medical dissecting room. I touch with fondness a mineralized skull vault whose age I can never prove but that rolled, I well know, for ages in the glacial gravels of the Platte. I look at them all, these silent masks whose teeth I have mended and whose mortal rags I have patched together with preservatives. ... Generally I can't refuse skulls that are offered to me. It is not that I am morbid, or a true collector, or that I need many of them in my work. It is just that, people being what they are, I know the skulls are safer with me. Call it a kind of respect for the bones, ingrained through long habit."

My own small collection is more prosaic. A plastic human skull, remnant of a "Visible Head" model I assembled in my junior high years, a dog, a cat, a possum, a mouse, the jaw bone of a cow, and a few petrified teeth that might be those of an ancient camel. I too am respectful of them, and not morbid.

Whale skulls, such as that of the grey whale whose image at the top of this chapter, don't call to mind those of other mammals because there are no eye sockets. At first glance that seems strange. But then one realizes that vision is a minor sense for cetaceans and eyes of minor consequence in the deeps.

Ulysses' less than enthusiastic relationship with North Carolina may have been mediated by the copperhead bite he sustained here in 1982. He came in one day with two puncture marks in the side of his face, which was visibly swelling as I examined him. I was glad for the power of that pancake engine as I jetted

Buckwheat fifteen miles up the twisted highway to Black Mountain, cutting curves and hammering the straight-aways. I felt like I was Mario Andretti.

This was the first time I'd dealt with a cat who'd mucked unsuccessfully with a poisonous snake. I was sure he was a goner and I imagined his only hope was a shot of antivenom. The veterinarian quickly disabused me of that notion.

"It was a copperhead," he said, "because with fang marks that far apart, if it had been a rattlesnake this little fellow would have died before you got here. If it was a bigger copperhead he'd probably have died too. But he's going to make it, and he's going to swell up like a balloon. You'll have to keep the infection drained, and give him an antibiotic." He told me it was very rare for any vet to administer antivenom to a companion animal, and he didn't know, offhand, where he would get it if he thought it was necessary. In the case of copperhead bites, he added, it was his impression that antivenom was rarely used even for people, since getting the dosage right was tetchy.

Later I'd learn that, contrary to folklore, there has never been a medically diagnosed human death from a copperhead bite. Over time I've come to regard them as beautiful and pacific critters. I once intervened as an old hillbilly took aim at a harmless black rat snake, and he insisted that it might turn out to be a copperhead.

"Herman, you know that's a blacksnake. And even if it were a copperhead, I see them in my garden all the time, and they just sit there and watch me walk by. They are real slow to bite."

"That's what makes 'em so dangerous," he said, putting his gun to his shoulder again. (The lucky snake had slithered away.)

Baby copperheads have a lime green nubbin on the tip of their noses which is thought to attract potential prey insects or small amphibians. I saw my first such

youngster when Ulysses walked into the house one day
and dropped it at my feet, intact and wiggling. The
banded body pattern was unmistakeable and I quickly
tossed the snake into a bucket for release further into the
woods.

An hour later Ulysses returned with a second
snake child. The next morning he brought me a third.

This was getting out of hand, so I made it my
business to follow him as he made his way to the rocky
top of the garden where he sat patiently, gazing at a large
flat rock. I then carried him back to the house, closed him
inside and returned to the knoll with a lidded bucket.

I grabbed the far edge of the stone and tipped it
toward me, a shield between my legs and whatever lay
beneath. There were two adult copperheads, one more
than three feet long—the bigger, stouter serpent was
clearly the mother. There were seven more infants. I
grabbed each adult by the tail and tossed them in the
bucket, then set about collecting the newborns and
adding them to the container. Finally I hiked about a mile
up a woods road to a more deserted out-cropping
(knowing that copperheads love to bask and that warmth
might be important for the youngsters) and deposited
them on the rocks.

Pit vipers on this continent include copperhead,
water moccasin (or cottonmouth) and numerous species
of rattlesnake. Pit viper eggs develop inside the mother's
body (unlike most other North American reptiles, which
lay their eggs in warm locations) and are "born" alive.
There is no aftercare, so my parent snakes weren't
hanging out to tend to the children, it was probably
simply a nice warm shelter and no reason to hurry off
after the big event. Even the presence of the male would
have been coincidental. (Among pit vipers, the sex of
adults is easy to observe because the female body is
significantly wider in front of the anal opening, then
abruptly narrow—a girth evolved to handle the writhing

mass of baby snakes delivered each spring.)

Tuxedo Joe would be bitten by copperheads three times during his fourteen year tenure, one of which put a fang cleanly through an ear, leaving Joe pierced. Each time his face and throat would swell with green-pussed infections that I drained with a hot cloth and treated with antibiotics. He also had a near miss with a rattlesnake.

One spring day I heard a buzzing that sounded like a cicada, but it was the wrong season so it piqued my curiosity. I scrambled down the steep slope below the house and spotted Tuxedo staring intently into the weeds where a large timber rattler was coiled, less than a foot from the cat, it's tail up and vibrating like mad. Tux was raising his paw, getting read to pop the snake, and I swooped in, grabbing him around the belly and jumping further downhill where I managed to land upright. Tuxedo struggled in my grip, not happy with my interruption of his game, and we watched the rattler, which was easily five feet long, slide off into the woods.

This aside into the relationship between cats and snakes is not remote from the broad theme of this book, concerning belief and its ramifications. Hatred of snakes is bound up in religion, with the serpent that purportedly tempted Eve an object of Christian vilification and widespread fear. Snakes are actually okay. Most are entirely harmless and the few poisonous varieties are easily avoided. They are an essential part of our ecosystems and absent serpents we are all much poorer, if for no other reason than the great service they provide in control of rodent populations.

Uniquely among the many cats who have shared my life, Tuxedo Joe apparently died of a stroke. One day he seemed to go blind in one eye and could only walk in circles, crying. The vet offered the diagnosis and said there was nothing to be done and that he might recover though she deemed it unlikely. In the event he lived another few days and had what must have been another stroke, convulsed and died. He's buried in the garden on Snookspeak.

Brave Ulysses' mother, Snook, made herself fully at home on that ridge in North Carolina—hence the name. She clearly loved that place, queen of all she surveyed from atop a huge bouldery knob at the crest of the garden. She'd lie belly-up in the sun or curled in the shade as the weather suited her, chase butterflies or ambush voles depending on her mood, and trot down the stone steps to greet us, mewing, when we returned home after hours away at work.

Cats' reputed nine lives are not wholly a fabrication. They show remarkable powers of recovery from accidents. Snook, for her part, was presumably asleep on my truck engine when I started it one afternoon. She screamed and tried to run off but was dragging both rear legs, the fan or fan belt having whacked her.

We collected her and raced those 15 miles to the nearest vet, who informed us that she apparently had a double sacral-ileal fracture—that is to say, her pelvis was broken off of her sacrum on both sides. The likelihood of her recovering use of her rear legs was exceedingly low and we might want to consider euthanasia.

The only possible hope was the veterinary school at the University of Georgia in Athens, so we phoned the school and blasted off for 175 mile drive. It turned out there wasn't much to be done, other than to help her keep off her legs, support her when she needed to use the litter box, and keep her warm and comfortable. With lots of TLC, she healed and recovered full use of her legs.

Months afterward Snook became listless, however, and a vet visit revealed that she was positive for Feline Leukemia. This was during the period when Fe-Leuk was just being recognized as a widespread infectious disease and vaccination against the disease was becoming available. We were informed that there was little hope she would survive and were advised to vaccinate the other cats.

Following a game plan offered in *Dr. Pitcairn's Complete Guide to Natural Health for Dogs and Cats* we began to provide her with vitamin supplements and fresh food (which meant the same for the rest of the feline team). Snook regained her health and flourished for another five years before going into an abrupt decline, refusal of all food, wasting and finally curling into a quiet bundle where she died in her sleep.

Like Tuxedo, Irma Burger the wonder dog, Azula-panula (euthanized after we found her in a tree, back broken by a raptor whose claw marks were all too evident), Streek (euthanized after inoculation with FeLeuk vaccine produced an unfortunate side effect— Feline Infectious Peritonitis which made her abdomen blow up like a balloon and cry out plaintively in pain), seven chickens and much later Pomonella (at 22 years, a very old cat), Snook became part of the garden at Snookspeak, pushing up an azalea bush year after year. As Thornton Wilder described it in *Our Town*, all are sleeping on the hill.

Doubtless some future gardener will turn up a skull or two somewhere down the line, skulls being among the most durable of animal bones. Nature is good about protecting our brains.

That hill taught me a great deal in my decades of up and down, back and forth. Never harvest firewood downhill from where you lay your fire—unless you have power equipment to lift it. Never expect to build up garden soil near a ridge—all of nature works against you. Think twice before removing natural predators (snakes for instance) from a garden plot—they eat more vermin than your assembled cats will achieve in their wildest dreams, at least if you feed the cats. Never underestimate the beneficial effect of a flock of free range chickens on an organic garden. Control your folly. And if two observations conflict, check your premises.

When we first began to clear the land I noticed a great many fallen locust tree trunks which were duly sawn for fire wood. (Locust is very dense wood and yields more BTUs per pound than just about any other wood in the mid-latitudes of North America.) Because the acre fronted on a road often used for logging, my first surmise was that the plethora of locust was a remainder from a logging operation. Or, on more reflection, perhaps it was timber downed by a recent big storm.

But soon enough, listening to farmers and sawyers in the valley, I came to understand that those logs wouldn't have been left behind by anyone with an eye for profit. Locust is the wood of choice for fence posts and while it has been supplanted by pressure treated lumber in many applications, it's still highly regarded for weather resistance (not to mention it's value as fire wood). The story goes that locust fence posts often last longer than the holes they're planted in, and farmers have to go out and pull them up and dig new holes when the old ones wear out.

What I came to understand was that the reason why so much of the downed timber on the hill was locust was precisely because it is so weatherproof. What I initially believed to be a spate of recent logging or windfall represented many years of falling trees. They rot very slowly.

The same is true of whales, particularly whales that settle to the uttermost bottom of the deep blue sea.

12. visions

O ne side makes you larger

In 1969 I was introduced to hallucinogens. That was clearly one of the great and enduring benefits of college life in that era. The subsequent devolution of drug use into simply getting messed up for the sake of getting messed up hadn't yet gained wide appeal. Or maybe it was just my circle of friends at Emory University and my own intellectual proclivity.

In my case, at any rate, I investigated the matter pretty thoroughly before I engaged, doing considerable research and even writing a college term paper on the subject of "technological innovation and the modern visionary experience." (It was a "B" paper, as I recall.) LSD was a product of the 20th century's increasingly technologized pharmacopeia, and was first synthesized in 1938 by Albert Hoffman at the Sandoz Laboratories in Basel, Switzerland.

In 1947 the drug was approved for psychiatric use, and in the 1950s the CIA began testing it on a wide variety of subjects, mostly without informed consent. When the Sandoz patent expired in 1963 the drug became more widely available and was classified as an Investigational New Drug by the U.S. Food and Drug Administration. Wide availability led to wide experimentation and it soon became a foundational drug for the growing counterculture. The anti-war and hedonistic life choices being made by the Woodstock generation frightened authorities. By the time I was able to obtain my first sample LSD had been illegal for a year and was therefore becoming even more widely available. Prohibition is ever thus.

As many other writers and artists, including Aldous Huxley, Timothy Leary, Baba Ram Dass (Richard Alpert), Alan Ginsberg and Albert Hubbard (to name a very few), have reported in the decades before and since, much of what one experiences under the influence of psychoactive drugs depends on one's mental frame. I dropped acid anticipating some sort of spiritual enlightenment and, sure enough, that's what I experienced. My exploration of cosmic consciousness was abetted in no small part by a young woman who was vending purple microdot LSD on campus and with whom I spent several rhapsodic dawns. We'd swallow the magic dots with coffee and greet the morning sun sitting in lotus position on a hillside not far from the college president's home. When the day brightened we'd read long sections of philosopher Alan Watts' books aloud, losing track of introductory sentences before reaching the distal ends of paragraphs, and grinning like jack-o-lanterns at the profundity of it all. Wherever you are, Beth, oh platonic spirit guide of my meager enlightenment, thanks for the memories. And, of course, the acid.

Mostly it was just wow!! accompanied by a strong sense that I was seeing some greater reality, some unity that encompassed all the great divisions of everything thingness. It was the 1960s, after all. Turn on, tune in, drop out and study war no more.

Having achieved the first two, I bought a gas mask and headed to Washington, DC, for the Moratorium Days demonstrations, hitching a ride with three Georgia Tech fellows (and my aforementioned spiritual advisor, Beth) for a Magical Mystery Tour that included a midnight appearance before a North Carolina traffic judge. The officer of the court shook us down for most of our cash-on-hand to spring our too-speedy driver. Fortunately he'd only dropped one of Beth's microdots and kept it together while dealing with the police officer and the court. After just a few hour delay we were back on the road again, laughing hysterically all the rest of the way to Our Nation's Capitol. Then it was tear gas and joyful rage as we demanded an end to the war, marching around the Pentagon and chanting.

"Ho, Ho, Ho Chi Minh, NLF is gonna win." (For younger readers, Ho was leader of the North Vietnamese government and the National Liberation Front included the North Vietnamese army and South Vietnamese who opposed the U. S. occupation of their homeland.)

We were later proved right, of course, though at the time it really strained my personal philosophy. I had never considered cheering for the other side in an American war and I had to go deep down in my thinking to decide whether to lend my voice. Opposing the war had been one thing, but that chant put it on the line. Up until then I had only considered our national war policy, and I wanted us out of Vietnam. But I hadn't much considered that our leaving necessarily meant that the NLF would "win." It was clear, once considered, that it did. And I was then willing to say it aloud.

One Tuesday night, in my dorm room, tripping on acid alone, I experienced a vast distancing, a sense that I could die right then and there and leave this earthly plane for something better. It seemed as if it were a conscious choice, and I teetered on the verge of the question. It was in no sense a suicidal impulse—that is, I wasn't depressed and had no intention of flinging myself out a window, cutting my wrists or finding rope and a stout limb. The notion was that I could simply fly inward to some other dimension. Or outward. It wasn't scary. It appeared to be a pleasant option as I listened to Graeme Edge of the Moody Blues.

Be it sight, sound, the smell, the touch.
There's something,
Inside that we need so much,
The sight of a touch, or the scent of a sound,
Or the strength of an arquebus deep in the
ground.
The wonder of flowers, to be covered, and then to
burst up,
Thru tarmac, to the sun again,
Or to fly to the sun without burning a wing,
To lie in the meadow and hear the grass sing,
To have all these things in our memories hoard,
And to use them,
To help us, to find ...
[Then the music crescendoed]
Ride, take a free ride, take this trip, it's for free ...

In the middle of that trip I received a phone call from my Mom, informing me that my best friend from grade school had died. He had fallen ill at his college, in central Illinois, the previous Thursday and had gone home for the weekend. He died on Monday of spinal meningitis. Bam! Doug's story was over. *Finis.* My trip turned a corner. Yes, I might be able to decide to die and follow that silver cosmic thread wherever it might lead, but here on earth there were others who cared. I was bound for and to home.

Perhaps this is a good place to interject that during some extensive experimentation with a few different hallucinogens I never felt any serious confusion about what was "real" and what was an induced perception other than that fanciful notion that I could somehow go inside and disappear from the earth. In the event I never tried that trick, and my supposition is that I would have failed if I had. But hey, you never know.

Recall the time when René Descartes walked into a tavern and the bartender said, "Would you like a beer?"

Descartes replied, "I think not," and disappeared.

There was always an element of perceived powerful connection between discontinuous events or ideas, the so-called "mind expansion" effect, a sense that karma was somehow playing out in real time. But the leaps of logic were clearly voluntary, there was never any serious doubt that the ecstatic sense of profundity and deep meaning was drug induced.

About three years later, in 1972, I was living in a cabin in the backwoods of Florida, adjacent to hundreds of acres of cow pastures. Cow pies of a certain age are host to hallucinogenic psilocybin mushrooms and the adjoining fields were ripe. My friends and I ingested substantial numbers of those fungal fruit during my time there, and experienced some amazing stretches of consciousness.

One experience stands out, insofar as I thought, once again, and with utter certainty, that I was happily experiencing and accepting death.

My then-wife and I ingested a good quantity of the psilocybes, the number not being of much significance. The potency of the 'shrooms varied due to weather and season, and there was no certain correlation between number and effect. In midsummer they might express as 5-inch-wide caps on seven inch stalks while in winter they shrunk down to a third in both dimensions. But the production of the alkaloid that triggered intoxication was apparently less associated with size than with heat, moisture and freshness. In any event, we cooked up a broth of the 'shrooms we had gathered, drank the bitter brew, and awaited enlightenment.

Thirty-eight years later the ensuing hours rank as the most powerful spiritual experience of my life and feel like my best theory of what dying will be like when I do actually expire. Or, at least I hope so. It's comforting to imagine that expiration will be thrilling, even enlivening if that isn't too contrary a notion. I was transubstantiated, rocketed, born into a new heaven on a new earth, lifted, exulted, graced, blessed and delivered. I saw worlds within worlds and heard the engines power up as the reality I knew was lifted off like a moon rocket and blasted toward the stars. There was a battle going on between the children of light (us included, of course) and the forces of darkness. Eternity beckoned. Everything mattered so much more than anything had ever mattered before. All of existence was up for grabs, easily in reach, ripe for the plucking, glowing with an inner fire and singing.

Oh yes, the world was awash in jubilant song.

It rained during that trip and in the aftermath the drops of water sliding off the leaves appeared as molten gold. We were baptized in the waters of that new day and the world appeared richer and dimensionality deeper. There was a sense of being able to reach into the energetic core of every living thing, plant and animal alike. Auras were visible and seemed palpable, a glow emanating well beyond the surface of skin and fur and bark and leaf and petal. There was a profound knowingness—looking out from some inner core of spirit and finding that spirit in others' eyes, animal and human. The feeling of unity with all of existence which I experienced has not left me in the four decades since, though it is now relegated to an operating assumption instead of whiz-bang mystical revelation.

There is nothing in the Biblical book of Revelations that seems more astonishing than the visions provided by hallucinogens, and by some lights that is exactly what John was about. "Locusts and honey" has been reported as a mistranslation of "mushrooms and honey," an inter-pretation given strength by the practice in some cultures of storing mushrooms *in* honey which is said to have an antiseptic and preservative effect.

If you somehow missed it, the world as we know it did not end in 1972, despite my earth-shaking experience. After that point we all continued to put our pants on one leg at a time. There was another working day to face, laying concrete blocks and mixing mortar, setting up scaffolding and calling in orders for redi-mix concrete—as well as trucks to repair, a garden to weed, goats and rabbits to feed, and on and on with the mundane details of homestead life. Outside a small circle of friends nobody even heard about our out-of-this-world adventure before this writing. Life went on.

What that experience confirmed for me is that all religious experience can easily be explained as a manifestation of personal consciousness, sometimes drug induced. Perhaps the reason there were more frequent reports of holy visions in the past was due to poor food handling, or the rye ergot fungus, and I'll discuss yet another possibility in a later chapter.

It is conceivable that we can communicate via telepathy, and certainly we will discover other conscious beings in the vast universe, given time, but we don't need the concept of God or gods to explain it all. On another front, consciousness might survive death, though I feel certain it does not. But again, such survival, in and of itself, would not require a god explanation.

It may be that one day we'll be able to communicate with the other big brained species on earth, the cetaceans, and discover an entirely alien and deeply enriching way of approaching reality—if we don't kill them all first, of course, either directly or by poisoning the oceans. That too will not require a supernatural source, any more than communication between humans of widely diverse backgrounds requires the same.

If someone says, as some do, that "all" is what they mean by "God," I'd say it begs the question. It isn't what most people seem to mean when using the term to describe a being that created the universe or inhabits objects or parts seas or stops the sun overhead or impregnates virgins or sits in stern judgment when we die.

Not too long after that powerful revelatory experience, following some milder but essentially similar trips, I let the psychedelic exploration go. I learned what I learned, and that was enough.

13. falls

Piccard dreamt the bathyscaphe.

His new deep-sea vehicle enabled human beings to study the ocean's depths in an entirely new way—up close and personal. The first bathyscaphe, a submarine designed to withstand the enormous water pressure miles below the surface, was contrived and constructed by Auguste Piccard in Belgium between 1946 and 1948. Propelled by battery-driven electric motors, the submersible regulated its buoyancy with the help of gasoline-filled tanks and iron ball ballast and came equipped with a powerful light to enable exploration of the dark ocean depths. In the unpoetic way of a scientist, Piccard named his creation the FNRS-2, a moniker unlikely to grace tee-shirts or bumper stickers.

Piccard's second bathyscaphe, dubbed Trieste after a seaport in northeastern Italy, was purchased by the United States Navy from that country in 1957. Three years later, Trieste, carrying Piccard's son Jacques and Lt. Don Walsh, dove to the deepest point on the Earth's surface, the Challenger Deep, in the Mariana Trench—a feat that has not yet been replicated.

New structural materials permitted further improvements in manned deep-diving craft. The submersible Alvin was commissioned in 1964 and soon made headlines—it was lost at sea in 1968 with its crew making a perilous escape. Recovered and refurbished it went on to vastly increase our understanding of the ocean deeps.

While some unmanned observation modules had spotted whale carcasses at great depth, it was an Alvin expedition in 1987 in which oceanographers first confirmed the phenomenon which became known as a whale fall. The discovery came only a decade after that of hydrothermal vents and the wealth of strange sulfide-eating life forms they support. Whale falls have since been observed by other scientists, by military submarines and have more recently been intentionally created for study purposes by dragging carcasses out to sea and submerging them in deep water. This places a whale fall at a known location that can then be studied over a long period of time.

When a whale expires in shallow water, its body is normally devoured by scavengers pretty quickly. Nature is a dedicated and tireless recycler. Deep water creates an entirely different ecological niche, however. At depths below 6,000 feet fewer scavenger species exist, and in the deepest deeps there is far less oxygen dissolved in the frigid water. Decay proceeds slowly and a massive carcass can provide nutrients for the almost otherworldlyy ecosystem for decades. Bacteria convert fats to sulfides and sulfide-eating life forms follow in their wake.

Most creatures at great depths subsist on what is known as "marine snow," the planktonic outfall from surface waters. Whale falls can provide as much organic material as a thousand years of marine snow, and a single skeleton can support more than forty thousand individual animals.

Bone-eating zombie worms and bristleworms, hagfish, crabs, prawns and shrimp, sea cucumbers, squat lobsters and the more familiar lobster of table fame, octopuses, clams, and even deep-sea sleeper sharks populate the depths. Large colonies of tubeworms often move in for the feast and some scientists now think that the giant tubeworms which populate volcanic vents may have evolved on whale remains. More than three dozen previously unknown species have been discovered at whale falls, among more than four hundred total, and methane-based life forms have been observed there as well, flourishing in the low-oxygen environment.

Nor is the survey anywhere close to complete, given that almost all of the research has been accomplished off the coast of Southern California. It stands to reason that varying conditions around the globe have given rise to as many or more distinct ecozones as have been observed on land. As I write this there are weekly reports of the discovery of new species on the falls, and it has been lately revealed that the carcasses of dead cattle also occasionally sink to the deeps and there support similar communities.

"Whale fall' has even come to refer to the whole communities of creatures that thrive among the sulphur-laden ooze of decaying whales. Just as windfalls deliver an unexpected wealth of ripened fruit for the taking or timber to be milled, the death of a whale delivers a host of nutrients to the sea floor. There can be up to two million grams of carbon in its blubber and oily bones, food for a host of creatures, some of which may be so specialized that they depend on dead whales to complete their life cycle. Because of that dependence it is probable that the heyday of commercial whaling spun off an extinction spasm in the deep sea as the number of dead carcasses radically diminished.

Detectable falls are relatively rare, though their number is presumably slowly increasing today since most commercial whaling was curtailed in the 1980s. Before that ban went into effect whale populations had been reduced by as much as 75 percent and wildlife population experts say that such a drop in a food source would result in a 30-40 percent decline in species.

Once located, falls are difficult to study, but

researchers are continually learning more about the new species to be found among the remains. Indeed, while it was once thought that the deep oceans were relatively bereft of life, the new view is that the ecology may be as varied and complex as that of tropical rain-forests.

Some scientists theorize that various deep-sea species may use whale falls as stepping-stones to expand their range and colonize other ecosystems, such as hydro-thermal vents and cold seeps. Given the randomness of whale deaths and the vagaries of ocean currents, carcasses must exist at many locations on the seabed, islands of nutrient in the ocean's bottom, with average spacings estimated at 16 miles throughout vast deeps which cover a full 60 percent of the earth's surface. By the same token, falls must be more common along migration routes and some researchers suggest an average separation of 4 to 10 miles is the more likely norm for places like the North American Pacific coast, frequented by humpbacks and greys.

As research has progressed it has been determined that some of the life forms found on whale falls today predate the emergence of whales. Fossils of the clams and mussels that populate whale skeletons are associated with the fossil remains of 200-million-year-old aquatic reptiles including ichthyosaurs and plesiosaurs.

But why is this important?

First, any information we can glean from whale falls concerning the deep sea is important because we know so little about that ecosystem. Understanding what goes on in the deep sea is critical because, although it is deep and dark and distant, it is still intimately linked to the rest of the world. For example, the ocean floor is the biggest sink for carbon dioxide and comprehension of the global carbon balance could prove to be important for analysis and response to climate change.

Carbon dioxide in ocean water is used as building

material by photosynthetic plankton which then die and settle to the depths as marine snow, or are eaten and move up the food chain before snowing down into the depths. Some scientists have suggested we fertilize the plankton with iron, which would boost their growth and thereby decrease atmospheric carbon dioxide. The major fly in that ointment is a counter-calculation which shows that if we add enough iron to make a meaningful difference in carbon we will toxify the deep sea by further reducing its oxygen level, killing off much of its life.

Moreover, the path of carbon that falls to the deep sea floor is largely a mystery and the question of whether it comes back to the surface is unanswered. Carbon is distributed by animals in the deep sea, partly due to whale falls and partly marine snow and other creatures, and it is thought that zombie worms and other deep-sea species that feed on the falls may play a significant role in the global carbon cycle.

Oxygen depletion might change the recycling of other nutrients as well. Over many centuries nutrients that fall into the deep sea are lifted back to the surface. Animals and bacteria that live in deep sea mud return them to the sea water, producing upwellings rich in nitrogen, phosphorus, and other nutrients that feed algae growing in warmer surface waters.

Whale falls support a wealth of other life forms for the same reason they have long been a target animal for humans: oil. Oil may constitute 60 percent or more of the weight of large whale bones, and the skeleton of an 85-ton whale contains as much as five tons of oil. Depending on its size, a whales' bones can contain enough oil to feed sulfur-loving species for close to a century.

There's another way that understanding the ecology of whale falls might provide useful tools to address climate change. While carbon dioxide is the most common greenhouse gas, methane holds in twenty-five times as much heat per unit of volume, so reduction of that gas in the upper atmosphere could offer an alternative means of curbing greenhouse warming.

Multiple life forms living on the falls consume methane gas. It has been suggested that injection of these or similar animalcules into the upper atmosphere might be a practical way to reduce warming, particularly if the heating triggers a runaway methane release from melting tundra. The dire threat of escalating temperatures and sea level rise could quickly overcome concerns about possible dangers of meddling with the atmosphere.

Whale oil was the first widely used liquid fuel and set the stage for liquid fossil fuels: kerosene derived from coal (later oil) and the rapid introduction of petroleum derivatives after the Pennsylvania oil strike of 1859. Petroleum arguably saved the great whales from extinction and smoothed the economic upheaval of post-Civil War emancipation as oil powered machinery picked up where the solar energy embodied in slaves dropped off. Steam powered traction engines found agricultural use in mid century and gave way to internal combustion versions in the early 20th.

The petroleum and coal economy blossomed and spread its polluting carbon around the globe until our emissions commenced to tip the climate toward catastrophic warming. Now a possible solution has been discovered on carcasses of the oily beasts who begat the system. It has all the appearance of a metaphoric Phoenix emerging from the carbonaceous bones of whales. New life comes from death and new ideas grow from old.

What we know and what we believe sometimes intertwine and sometimes strain in opposite directions. When they conflict it is well to remember that belief is an unreliable pilot as we thread the shoals of reality.

ice

| awoke with a gasp.

My upper body was soaked and suddenly frigid. "What the ...?" was all I could manage by way of inquiry into the circumstances of the inundation. Taking a deep breath, I opened my eyes while my arms twitched amidst rattling ice cubes. Chill water spread around and beneath me, soaking into the mattress as I gathered myself toward coherence.

"Susan ? Why?" I asked, quietly.

Susan was nowhere to be seen. The stainless pitcher with which she had delivered up her anointing fluid clattered to the floor as I moved sideways off the bed. "Susan?" I queried more forcefully.

"Bastard!" rolled up the steps.

I stepped to the door and called out again, a single word that asked two questions. "Susan?"

"Teach you to fall asleep when I'm talking," came the reply.

What the hell had she been talking about, I wondered, plowing backward through sleep in search of some urgent subject. While I tried to recover our conversation I stripped the bed, mounded the damp bedding and spread two towels over the widening dark stain. Sleep beckoned, but I knew it was no use. She'd only wake me again with whatever. This was about twelve years into a relationship that lasted twenty-four. It wasn't the first time I was doused with cold water in my sleep, nor would it be the last. But it was the first and only time when she combined it with ice cubes. Nice touch, that.

I grabbed a spare blanket, propped the drier of the two pillows against the headboard, stretched out along the edge of the mattress and settled back, pulling a book from the stack beside the bed. Might as well read. Sleep did not appear to be an immediate option. There was no sound from downstairs and I eventually drifted off.

When I awoke the next morning the house was quiet. I glanced at the clock to find I'd overslept by an hour and threw on clothes, scurried downstairs and found a pink sticky on the fridge. "Later." was all it offered, over a scrawled "SKM." My keys were missing from my pants pocket and I ran back upstairs to check the dresser. Then out to my pick-up, locked tight, keys on the dashboard beside another sticky note. "Bastard."

I damned her temper and her twisted sense of humor and went around to the back of the truck, opened the cap and the tailgate and commenced to unload eight sheets of drywall, tools and two saw horses, crawled in and slid open windows in the topper and the cab and stretched forward to grab the keys. Reloading took longer. "Damn, damn, damn," my muttered mantra while I wondered again what she'd been telling me when I'd fallen asleep.

Then I was on the road and thinking about the current construction job, already two weeks behind. "Thanks," I fumed. "Thanks." And in just another hour or so, I forgot the whole matter, tucking it under the ongoing blanket of denial that made life passably comfortable.

It would be many more years before I began to sort out the wreckage of primal relationships, abuse, martyrdom, and deep pain on both sides of my longest-running partnership.

14. diet

Most of the time we don't eat our own.

What we term "higher" species generally exhibit some inhibition in regard to consumption of relatives. Guppies may eat baby guppies and frogs ingest incautious froglets but by the time life moved up to warm bloodedness the inclination to consume progeny was greatly restricted. Among humans, cannibalism has always been the exception, and generally limited to eating enemies (usually with a generous dose of ritual significance) or in extremis, where survival is at stake and then—at least as customarily reported by the diners—only after the dietary object had expired. Plane crashes, marooned arctic explorers and shipwrecked sailors in life boats come easily to mind. It's simple enough to grasp that any species which developed a strong culinary preference for its own children would be self-limiting, particularly a species like ourselves with few offspring produced at wide intervals.

In New Guinea and elsewhere, kuru, a neurological disease, proved common among cannibals and Creutzfeldt–Jakob disease, otherwise known as Mad Cow, emanates from the same source. Mother Nature seems to know what's good for us, and what isn't, regardless of the economics of animal husbandry.

A growing body of evidence—actually a growing body of skeletons—has led anthropologists to conclude that the Anasazi were cannibals. This has come as something of a surprise, since their descendants—the Hopi, the Zuñi and other Pueblo tribes in the American west—are among the most peaceful cultures in the world. It has been long and widely assumed that the builders of the great cliff dwellings and road systems in the Four Corners region were similarly pacific.

That assumption is greatly at odds with the discovery of baked skulls from which cooked brains have been eaten, and splintered long bones devoid of marrow. Add to that the presence of human grease on the interior of cookpot shards, and human remains in human coprolites (the polite scientific name for fossilized poop), and—well—you get the picture.

Death fascinates and repels—as it must for we who endure, obvious beneficiaries of a powerful and ancient survival instinct—and cannibalism is right out there on the edge. Whether it is a thousand year old Anasazi roasting pit piled with the butchered bones of men, women and babies, or a freezer in Wisconsin where a wacko stashed his sexual (dietary?) conquests, the evidence of cannibalism makes us squirm.

In warfare we may make allowances for so-called "collateral damage," but the death of innocents is more generally called murder and one of the chief demarcations between those we deem civilized and those who are not involves treatment of non-combatants. In the Vietnam war it was the My Lai massacre that most repelled. In Rwanda it was mothers wielding machetes to hack enemies' babies to death that shocked the world. Ugandan ex-dictator Idi Amin is said to have eaten children, and thus placed himself beyond the pale of civilization. When Afghan wedding parties are blasted to bits by an errant missile from a U.S. fighter aircraft , the world recoils.

We bring those sensibilities to bear on murderers as well, hence the enduring debate about capital punishment. Yea or nay may get the upper hand for a historical while and then shift again, though the trend line seems to be running against execution. But whether we opt for execution or life in a cage, we reserve our most extreme punishment for those who brutalize the innocent. At the same time, we offer grace to those who do our bidding.

One member of a firing squad is supplied with a blank bullet—all together they aim and fire, but each may be innocent. Electrocution or poison gas allow a little personal distance between cause and effect. Lethal injection is softened by the supposition that it provides the gentlest entrance to that final goodnight. In days of yore the axe man wore a hood to the chopping block.

The executioner is granted absolution because the sentence is imposed by others and the condemned is declared to be non-innocent. When that is unclear, we waver. Recall the photo, again from Vietnam, of a man with head rocked sideways as a pistol bullet exits his skull: guilty or no? Beyond our opinion of that war or any war, we are the more deeply troubled by doubt. Who was that prisoner? What did he do? Whose finger on the trigger? By what authority?

More recently, the execution of Cameron Todd Willingham in Texas in 2004 raised a storm of controversy. Willingham was convicted of murdering his own children, based on the testimony of a self-proclaimed arson expert. But that expert's track record suggested a strong proclivity for detecting arson, regardless of the facts.

Other experts have demonstrated that Willingham was probably telling the truth, that the fire that spread through his home on that fateful night was, in fact, caused by a faulty heater. Willingham insisted upon his innocence right up to his death. We killed him anyway and Governor Rick Perry, readying a run for re-election at the time, decided to block an official inquiry. A devotee of the death penalty, Perry obviously thought that a deeply flawed process didn't bear close scrutiny, particularly given the number of potential Lone Star voters who seemed to favor execution.

Considering the large number of condemned murderers whose convictions have been overturned due to DNA evidence, proof of police or prosecutorial misconduct, or insufficiency of counsel, one needs to be buried in denial to believe we haven't executed many innocents through the years. Furthermore, investigative operations like the Innocence Project have very limited resources and so restrict their work to convicts headed for capital punishment. It stands to reason that there must be at least as high a percentage of those meted out lesser sentences who have been similarly misjudged.

It's no wonder that the death penalty has been eliminated in most modern democracies. But vengeance has its adherents, especially among those who adhere to faith traditions that claim vengeance is the sole province of the Lord. Since the falsely convicted are often drawn from the criminal fringe, many people in this country seem to reason that we are well rid of most those executed, despite the occasional mistake. The general squeamishness about eating our own is more broadly accepted, but even that reticence can be jettisoned in light of perceived good.

When I was a child I couldn't get enough of the Hall of Egypt at the magnificent Field Museum of Natural History in Chicago. Amidst all manner of relics from a strange and distant civilization lay the mummies. Human prunes in glass cases were scary and magical and weird all at once. And they all appeared to be grinning! At age eight I was already gripped by the tangibility of another human face gazing down through the centuries; an experience which seems to resonate with most of our species across cultural and ethnic divides.

Mummies have been guaranteed crowd pleasers from the time of those ancient Egyptian funerals with paid mourners and lavish burial feasts right up to the present. Since 1911 when three silent efforts appeared in France, Britain and the U.S., dozens of major mummy movies have been released with countless translations, variations, imitations and clones around the globe. (More about the clones later.) It is easy to get the sense that directors just love the chance to shout "That's a wrap!" with their main character done up in shredded sheets.

Most critics agree that the 1932 Universal Studios release starring Boris Karloff is the best of the lot: later versions were merely attempts to milk the cash cow. Karloff's masterpiece, filmed in three weeks (with make-up consuming eight hours of every shooting day) was a triumph. As Egyptologist Bob Brier has observed, "The reason for the success of this fictional treatment above all others is the humanity of the mummy. (He) ... has a full range of emotions—he lives, fears, and gets angry. He is the lover desperately seeking to be reunited with his love."

While wrapping actors has been the rage in the cinema century, unwrapping was the big draw for a hundred years before that. As travel to Egypt became relatively affordable in the 1800s, mummy collecting blossomed. Egyptologists drew overflow crowds to lectures which culminated in unwrapping a cadaver. With scissors and knives, sometimes with chisels and saws (old resins can be stout), and always with dramatic flourish, the experts would reveal the windings, the amulets, the jewelry and finally the naked torso of a 3,000-year-old decedent.

People being people, exposure of the groin always seemed to excite particular interest. But, dust tending to return to dust (as it will), the soft tissues of mummies don't cohere after a few hundred centuries: despite the high excitement there wasn't generally much to see.

Before unwrapping hit the big time, grinding was the ticket. It seems that ancient Persian texts referred to "mummia" as a wonder drug—the Prozac or Claritin of its day—a concoction of magical power only available to the lucky few. Along the way some translator goofed, applying "mummia" to the balsam resin used to soak Egyptian funerary linen ("balsam" and "embalm" share etymologic roots), usage meandered until the term included the body inside the fajita wrap, and a booming business emerged. Whole mummies were ground to powder and sold as nostrums throughout Europe. (Note that quite apart from the big-name royal cadavers of Pharaohs and their wives, there were thousands of everyday mummies available as grist for the snake-oil vendors and medicine shows.)

(What is it about people and powders anyway? Pulverized mummies, BC, cocaine, snuff: grind it and they will come. Made more puzzling by the phrase, "Take a powder," which, curiously, means "to go.")

In any event, Europeans circa 1700, were sold on the benefits of eating dead Egyptians, as a tonic. Leeching, one notes, was also all the rage, creating a sort of eat or be eaten approach to optimal health.

The reason mummies were mummified, as everyone knows, emerged from the Egyptian religion. Like many of today's devout Hebrews and Catholics, Egyptians believed in literal resurrection and that the decedent's body would be reused in the afterlife. Being somewhat more pragmatic than other cultures, and also having long experience with salting fish, the Egyptians came up with pickling as a great way to keep the mortal vessel ready for its next voyage. ("We've got your pickle, it's a pickle named Hatshepsut.") They even went to the trouble of saving all of the important organs in separate containers with seven different herbs and spices to insure maximum readiness for heaven.

But they didn't save the brain.

Ever.

Think about that for moment. You almost certainly believe that your "self" is centered in your head. That idea is deeply ingrained in our culture and well supported by neurological and biological research. But the Egyptian civilization was hugely successful for a very long time while embracing a completely contrary belief.

The brain wasn't considered an important item since it was "known" that the heart ran the whole operation. The brain was more of an inconvenient puddle which needed to be drained from the departed's head and replaced with hot resin. No extant hieroglyphic text has been found which links the brain to any important function in the body. Notwithstanding three or four hundred centuries spent beating their foes and each other over the head with stone and bronze blades, they never quite got the drift, and it may be that they weren't actually conscious in the sense that you and I would use the word.

So, like most of the films on the subject, mummies are brainless.

Brainless, but not clueless.

By studying mummies—first with the medical technology of the 19th Century, later with electron microscopes, x-rays, CT-scans, and now through sampling DNA—scientists have opened a window on everyday life forty centuries in the past. The biggest surprise may be that there isn't much to be surprised about. Intestinal parasites were a problem then as they are now in that region. So too was silicosis caused (then as now) by breathing excessively sandy air. Elderly pharaohs were prey to hardening of the arteries and arthritis. Most of them had very bad teeth due in large part to the use of sand in milling grain which then eroded dental enamel. In fact, infected teeth were a major cause of death since they had no antibiotics. (This puts an interesting spin on the old phrase "the bread of life," does it not?)

Forensic pathologists have even pretty much buried the long running rumor that Tutankhamen was murdered, having found evidence that he was a genetic cripple due to inbreeding, and probably died due to complications from malaria.

Scientific curiosity, tending as it does to leave no Pandora's box unopened, has even led to successful cloning of mummy DNA, which some boosters enthusiastically predict will lead to cloning of a whole ancient human.

Picture this: One day when a young woman has come of age her adoptive parents tell her they need to have a little talk.

"Is it my grades?"

"No, Hattie, your grades are fine. We just need to tell you some facts about your life."

"Thanks, but, no thanks. I had that pretty well figured out in the 6th grade. And, don't worry, I know about STDs."

"Honey, would you just listen? We've never hidden the fact that you were adopted ..."

"I knew that from my nose. Do you think I could get a nose-job for graduation? That would be so cool."

"Don't interrupt.

"You are a clone. You are Hatshepsut, one of the four women who were ever crowned Pharaoh during the reign of the longest running monarchy in the history of the world. You are a King of the Eighteenth Dynasty, which expanded the Egyptian empire to its furthest extent. You ruled Syria, Iraq, and most of North Africa."

"Mom, Dad, are you on something?"

"What?"

"I mean, I know you probably smoked pot in college. Have you started up again?"

"Hattie! This is serious." Then, taking out a notebook, "See, its all here in this file. You are Hatshepsut."

"Right. Is that all? Gotta run, soccer game."

Later would come the law suits. Go figure. If Jews can recover treasures stolen two generations ago by the Nazis, and Native Americans can recover artifacts stolen six generations ago by European invaders, won't the courts have to award Hattie her due after a thousand generations of theft? Museums around the globe will be emptied of their Sphinxes, their obelisks, their coffins and their gems. King Tut's gold alone would make her one of the richest people on earth.

What then? A captain of industry? A personal army? A Senate seat in New York?

And finally, this: Hattie staring down at a crumpled collection of leather and bones. A pickle. A prune. Herself. Once again physically alive, but with the meaning gone. A life out of context. A deep sense of irretrievable loss.

Sic transit gloria mundi. The mummy has returned.

But, I digress. (Mummies are way too cool.)

The point I was making is that who or what we eat may have begun as a function of evolutionary survival but has come to be a matter of belief. So we eat or don't eat things based on a little diet science, a lot of cultural suggestion and availability. During the 99 percent of our history that we were gatherer-hunters, members of our species living near the equator ate more vegetables than those living nearer to the poles. As areas became more crowded the average size of animals consumed shrank. As land animals became scarce, oceanic sources became more important. When agriculture was invented, cities became possible and the variety in our diets began to dwindle based on what could be most successfully grown or husbanded. In modern times, various religions began to proscribe or prescribe foods, and more recently we have begun to apply relatively abstract ethical precepts.

Few modern westerners are likely to be comfortable eating something with a face resembling our own. There is reportedly a substantial market for "bush meat" which is the standard euphemism for chimpanzee and bonobo flesh when it travels from the poaching ground in Africa to ex-pat African diners in Europe. But those ex-pats are presumably pretty recent émigrés and mutilated monkey meat* was what Mom served them. The attraction will probably fade as they assimilate. Similarly, there are stories of Japanese epicures who fancy eating the brains of monkeys served alive, but one has to wonder how prevalent that might actually be, in practice.

On the flip side, gorilla hands and feet are regarded as a delicacy in some parts of Africa.

There must be vanishingly few modern Westerners who would be comfortable eating human-like hands whether from a chimp, a gorilla or a monkey. They are too close to us on the family tree. Chicken fingers, yes. Monkey fingers? I don't think so. In that regard there is a growing distaste for using our near relatives for medical experiments, though it's a bit simpler to ignore the plight of caged animals kept well out of sight while they are infected with ebola, immunodeficiency viruses and worse, than to consider something on your plate.

Looking backward, there aren't widely accepted reports of *homo sapiens sapiens* dining on Neanderthals during the thousands of years when our ranges overlapped. So perhaps the taboo against eating upright bipedal near-relatives is tens or hundreds of thousands of years in the making. And yet, modern gatherer-hunter

***

*From the childhood jingle:
"Great green gobs of greasy grimy gopher guts
Mutilated monkey meat, dirty little birdies' feet
French fried eyeballs floating in a pool of blood
and me without a spoon."*

groups continue to consume apes and monkeys, so that narrow distinction is ancient and enduring as well. For that matter, Jane Goodall observed chimps eating baby baboons, another evidence of near-species carnivory.

Outside of China, dogs are almost universally rejected as lunch meat (except during famine) and only the Chinese and Vietnamese eat cats as a regular dietary item; Hindus, some Chinese and many Zoroastrians don't eat cattle; orthodox Jews and Seventh Day Adventists don't consume pigs; Jews also abhor horses, bears, elephants, camels, amphibians or reptiles some birds of prey and bats, shellfish, freshwater eel, catfish, most insects and rabbits; Muslims share the proscription on pigs, horses, and predatory animals generally, while Shia Muslims add eel and rabbit to the no-fry list and Sunni Muslims eat rabbit but reject donkey; most Somali clans and most Cushitic speakers eschew fish (Cush is spoken in parts of southeastern Egypt, Ethiopia, Eritrea, Somalia, Kenya and northern Tanzania); Navaho, Apache, Zuñi and some other southwest American tribes have a taboo against fish and other water-related animals, including waterfowl; Krishnas abjure fungi and plants of the onion family; residents of the Balkans and Poland won't eat horse meat, and horse is an unusual menu item outside of Japan, France, Germany, and Kazakhstan; rats and mice are off most grocery lists, with exceptions for certain species in some areas of Southeast Asia and Africa; Yazidi Kurds won't eat lettuce; and Australian aborigines have personal or clan totemic animals which are proscribed.

Broadly speaking, our modern Western food ethic comes increasingly to rely on the Golden Rule in regard to much of the living world. There is a growing belief that food animals should not be made to suffer. The widespread revulsion exhibited over football player Michael Vick's recent participation in dog fighting reflects the widening sensibility that tormenting animals for amusement is no longer acceptable. We have come a long way from René Descartes who drew a clear distinction between animal and human, insisting that his dog felt nothing when he kicked her and she yelped.

Descartes' scientific peers went a step further and registered no compunction about nailing a dogs feet to the floor or performing bizarre and painful vivisections sans anesthesia. Such torment would find very few supporters in the scientific community today.

Another track in our extension of personhood is the recognition of intelligence and compassion. It's difficult to imagine, if we are ever visited by aliens from some distant world, that we would treat them much differently than we do other human beings—assuming that they came in peace, of course. Their relative smarts would be beyond question, insofar as they got here before we got there. The likelihood that we would consider them food animals, even assuming we could overpower their superior technology, seems even more remote.

But that conjecture may depend more on their appearance than on their mental acuity. If they are bipedal and look us in the eye they will certainly get a better reception than if they resemble clams or slugs or serpents. It may be difficult to imagine a technological civilization sans opposable thumbs (or the manipulative equivalent), but that may simply reflect a failure of our imagination.

Cetaceans in particular have been elevated by the comparison of brain size and complexity. There has been an enormous amount of research done on bottle-nose dolphins and their intelligence is beyond question. Of course their brains are adapted to very different living conditions and interspecial communication is necessarily limited by that disparity of experience.

At the same time, numerous stories of swimmers or sailors saved by dolphins make it clear that they share our sense of altruism and a distinct awareness that we have different needs that aren't well served in mid-ocean. Bottle-nose dolphins are closest to humans in their brain to body weight ratio, a similarity which may have some bearing on what we are able to recognize as intelligence. On the other hand, elephants sport brains five times the size of ours and the intelligence of elephants is widely acknowledged. Ethologists report that Asian elephants, great apes, bottle-nose dolphins and magpies all exhibit some measure of self awareness. Still and all, humans in Africa eat elephants and apes and the Japanese eat dolphins.

Whales possess the largest physical brains of any animal but there is no scientific consensus about the existence, nature and magnitude of their intelligence. In part this stems from the difficulty of studying the behavior patterns of large oceanic creatures. Humpback whales have been found to have spindle neurons, a type of brain cell previously considered to exist only in dolphins, humans and other primates and it is well established that some species of whale are highly social.

We don't understand whale song but we know that they sing. We can't comprehend the mental framework of hundred ton beings who evolved in virtual weightlessness with no reference to "feet on the ground" or right angle geometry or tool-making or agriculture or division of labor; beings attuned to a sonic environment far more complex than anything we experience, and possessed of an alien wisdom expressed without written records or the help of opposable thumbs. Most humans who have spent even a little time considering the cetacean mind are enthralled by its possibilities and otherness.

Once there was a way to regularize the availability of whale products on a profitable basis, European culture had no reservations about categorizing whales as a perfectly suitable industrial feedstock. When that mind-set spread around the globe it drove the great whales toward extinction. Long after petroleum mostly supplanted whale oil and baleen, the slaughter continued, fueling continuation of subsidiary uses including steaks in Japan and mink farming in Norway.

But other people approached whale slaughter at a subsistence level, and those included the Makah in North America's Pacific Northwest.

15. indigenes

The Makah kill whales, or try.

For a hundred or more generations, the Makah (or "People of the cape") inhabited a large portion of Washington's Olympic Peninsula, extending from Cape Flattery at the northern tip of the peninsula for many miles south along the Pacific coast and east along the Strait of Juan de Fuca. Archeological research has documented well over two thousand years of Makah life and culture at the village of Ozette, about fifteen miles south of Cape Flattery.

While hiking along that rugged coast I've seen the cartoon-like petroglyphs etched into the stone faces of promontories—carvings more reminiscent of modern graffiti than the cave paintings of Lascaux or Altimira. Not that this observation is intended to demean the artwork. It was simply the case that when I came across said petroglyphs while hiking, my reaction was instant disgust that some recent visitor to the Olympic National Park had felt called upon to deface the pristine trail-side surfaces. Only later, based on archeological documentation, did I come to understand that the crude artwork I'd encountered was actually centuries old.

Makah folk tradition suggests that they have always hunted whales, most particularly the humpback and grey whose migration routes run just offshore. Paddling cedar canoes, carved from the trunks of single trees, which held eight or ten men, Makah are said to have hunted and fished far out to sea, sometimes more than one hundred miles. Early white observers commented on the Makahs' great skill as canoeists and as whale hunters.

Consequent to that tradition which is enshrined in the Treaty of Neah Bay, the Makah are the only tribe in the United States with a legally guaranteed right to hunt whales. But in 1855, the same year the Makah signed their accord, a New England whaler named Charles Scammon discovered the greys' birthing lagoons in Baja California. Commercial whalers flocked to the lagoons, and within a very short time the grey whale was nearly extinct, its population dropping from perhaps 30,000 to less than a tenth of that number.

Despite their treaty rights, and in part due to the collapse of the whale population, the Makah voluntarily abandoned whale hunting for most of the next thirty years in favor of the lucrative commercial fur seal trade. When the fur seal population had been almost completely obliterated, at the end of the 19th century, the federal government finally moved to halt that slaughter. Many Makah hunters returned to hunting whales on a limited basis—limited due to the sharp decline in target animals. The Makah sporadically hunted and traded whale products until 1915, held a few final hunts in the mid-1920s, and ultimately quit.

Grey whale numbers rebounded following a general moratorium imposed by the International Whaling Commission in the 1960s. By the 90s the grey population was theoretically high enough to qualify for removal from the Endangered Species List. In the mid-90s the tribe announced it would resume whaling in order to revitalize its putatively moribund culture, a decision which ignited worldwide controversy. While animal rights activists bitterly denounced the Makah, other groups, from advocates for indigenous rights to the

United States government, defended the tribe's right to hunt.

According to tribal folklore, traditional whaling started with complicated rituals. Prior to the hunt, Makah tribesman would ritually bathe in the icy waters of the Pacific, then rub their skin raw on sharp mussels and barnacles. A few days before their hunt they would often dig up a fresh grave and dismember a human corpse. During the actual hunt they would lash the torso of the corpse on their backs—a sign of respect for their departed comrade. During the entirety of the hunt the wives of the hunters were required to remain motionless in their beds, not eating, sleeping or talking.

What the Makah did not choose to stress in their announced revitalization effort is the decidedly mercantile nature of their tradition. Unlike the subsistence hunters in many indigenous cultures, the Makah had engaged in active trade of whale meat, as well as fish, seal, and other sea-derived products with other tribes. This had segued naturally into trade with Europeans when they began to arrive in the 1700s. The Makah aggressively traded whale meat and oil through the mid 1800s.

So when the Makah signed that treaty with Washington territorial governor Isaac Stevens in 1855, they were seeking protection for the right to commercial exploitation of humpback and grey whales, not preservation of a spiritual practice or ancient tradition. And while the Treaty of Neah Bay is the only Native American treaty that explicitly granted a tribe the right to hunt whales, it also forbade them from trading whale meat internationally—a telling exclusion.

While advancing arguments for resumption of whaling, the Makah failed to make mention of instigation by the Japanese. In the midst of extended legal wrangling a 1995 memo written by Mike Tillman, Deputy Commissioner of the U.S. Delegation on Whaling Issues, came to light. He reported that both Japan and Norway had contacted the Makah about purchasing whale meat, and the Makah were considering construction of a processing plant. In addition, it was revealed that the tribe had signed a contract with a Japanese seafood company subsidiary in Alaska. The parent company is alleged by critics to have been repeatedly implicated in trade of black market whale products.

Thus the public relations face of the return to whaling was painted in terms of tradition and preservation of indigenous culture, yet there were big dollar signs hovering in the background. The Makah insisted that they had no intent to sell whale products other than whale bone carvings produced by native artisans. In 1999 Japanese market prices pegged the value of one grey whale at anywhere from $500,000 to one million dollars, and given that the Makah were the only Native Americans with a legal treaty right to hunt grey whales, they would have no competition. Moreover, if the assertion of traditional rights successfully trumped the IWC restrictions, Japan was clearly ready to push the issue on a global scale.

Whale hunting was prohibited by both the U.S. Endangered Species Act, which protected humpback and grey whales, and the International Whaling Commission, which had imposed that global moratorium banning all commercial whaling as of 1986. Whatever their motivation, the Makah went after the Endangered Species listing first. While humpback whale populations remained ominously low, grey whale numbers had rebounded. From a nadir of as few as 4,000 in the sixties, the world grey whale population had grown to as many as 22,000 whales. The Makah appealed to the U.S. Government to have the grey whale removed from the Endangered Species list and in 1994 the U.S. Government helpfully complied.

The Makah went into high gear, training a team of whale hunters, building a suitable sea-going canoe and issuing a steady stream of press releases. Following heated legal battles and physical confrontations with protestors, Makah whalers landed their first whale in more than seventy years on May 17, 1999.

While the Makah relied on the old treaty and claims concerning tradition in their assertion of rights, they relied extensively on modern technology, including stainless steel harpoons and high powered rifles—the rifles described as a humane addendum, used to ensure that the whale died quickly. They also relied on electronic communications gear, internal combustion motors on support vessels (though not on the hunting canoe itself) and an observation helicopter which somewhat calls into question their traditionalist claims.

On the one hand it can't be denied that the Makah culture was irreversibly damaged by the intrusion of European interlopers. But to the extent that they have availed themselves of that culture's benefits and technologies, have they subsumed their history to a newer paradigm? When you offer prayers to traditional gods and paddle a traditional dugout canoe but chat on cell phones and fire a high powered rifle, exactly whose traditions are you following? At what point are you required to choose?

The Treaty of Neah Bay expressly guarantees that: "The right of taking fish and of whaling or sealing at usual and accustomed grounds and stations is further secured to said [Makah] Indians in common with all citizens of the United States."

In a statement opposing Makah whaling, the Sea Shepherd Society wrote:

> The treaty that the Makah cite as evidence of their right to whale specifically states that they have the right to whale 'in common with the people of the United States.' When the treaty was signed, all Americans had the right to kill whales.

When whaling was outlawed for all Americans it included the Makah as the rights are 'in common' and not separate. There cannot be unequal rights granted in a system that promotes equality under the law. This is tantamount to extra special rights for a group of people based on race and/or culture and is contrary to the guarantee of equality under the law as guaranteed by the U.S. Constitution.

In the old days virtually every part of the whale was used. The oil, blubber, and flesh were eaten, sinews were used for ropes, cords and bowstrings, and the stomach and intestines were dried and inflated to hold oil. Even the bones were occasionally used in house construction. In the early years of European contact the Makah frequently produced a surplus of whale oil and blubber, which they traded to other tribes, and then to white settlers when they arrived.

In the 1999 hunt they speared and shot to death a young grey whale and dragged it to shore amidst a cacophony of opponents both physically present and in the courts. The carcass was then prayed over, as were the several whalers. Prayers were offered to thank the whale for giving its life to sustain that of the Makah and to free its spirit for passage to the other side. "Giving" in this context is a euphemism, of course. The whale was not consulted in advance.

The carcass was carved on the beach in preparation for a potlatch feast. Makahs of all ages reportedly ate fresh blubber, most for the first time. However, grey whale is not considered to be particularly palatable and was not much used for human food in the past—the flesh, together with the blubber and bones, was rendered for oil. Subsequent reports indicate that most of the meat from the 1999 kill was ultimately thrown away.

The potlatch, attended by members of native groups from the northwest and around the world, was held the following weekend to celebrate the successful hunt as a rebirth of Makah culture and a victory for treaty rights of all indigenous peoples. In fact it would eventually prove to raise a significant threat to those rights.

Protestors quickly condemned the whale's death and the Makah celebration. A candlelight vigil was held in Seattle and newspapers throughout the state were deluged with letters and e-mails denouncing the hunt and the tribe. Normally pacific animal rights activists were so outraged that some reportedly uttered death threats against the Makah while other comments took on a racist slant.

Repeated efforts at further whale slaughter were unsuccessful in early 2000, but Makah leaders insisted they saw lasting positive effects from the successful hunt. Interest in language classes increased, possibly ensuring survival of the previously diminished Makah tongue for another generation and high school students assembled the bones of the whale for display in the Makah Cultural and Research Center.

In years since then the issue has been the contentious subject of court battles and regulatory decision making. Moreover, in a ruling seen as having sweeping implications for all Indian treaty rights, a three judge appellate panel announced that the hunt cannot proceed unless Makah whalers obtain a permit or exemption under the Marine Mammal Protection Act

The Makah, other Indians, and experts in Indian law were all stunned by the ruling that the Marine Mammal Protection Act applied despite the treaty guarantee. Legal experts opined that the ruling was in conflict with the long-standing principle that Indian treaties cannot be overridden by general statutes.

The tribe tried twice to convince the appeals court to change its ruling, but to no avail. Advised that an appeal to the U.S. Supreme Court would risk further weakening of treaty rights, tribal leaders decided it was safer to comply with the court ruling. The tribe submitted a formal request to the National Marine Fisheries Service for a waiver of the MMPA and began work on the full environmental impact statement.

As the years passed with no authorization for a hunt, some Makah lost patience with what they considered a violation of their treaty rights. In September, 2007, five whalers again harpooned a grey whale and killed it with a large caliber rifle. The whale died within twelve hours, and sank after being confiscated and cut loose by the United States Coast Guard. This ensuing condemnation from anti-whaling forces was echoed by tribal leaders, who realized that the unauthorized hunt would undermine efforts to obtain the MMPA waiver. The whalers were ultimately convicted in federal court.

Today the Makah continue to insist it is imperative that they hunt whales to save their culture from destruction and continue to challenge the IWC ban. If they succeed in overturning the moratorium, the Makah may very well open a floodgate as others demand recognition of "traditional" rights. That could easily result in the extinction of most of the world's whales.

The songs of the whales would then go missing, before we ever had a chance to understand the music.

16. mind

What you eat you are.

Perhaps more to the point for human beings, most of us probably decide what to chew or eschew based in significant part on what we think we are.

Those of us who subscribe to some form of deep ecology are disinclined to draw sharp lines between species or between the often divided domains of animal/vegetable/mineral: Recall Vernadsky's description of life as a "disperse of rock." But most everyone subscribes to some form of hierarchy in the natural world. The most critical and therefore difficult choices often seem to reflect ideas of consciousness or feeling.

Consciousness is an attribute of living beings, and few among us credit plants with having it at all, though some may beg to differ. In any event the possible thought-processes of a potato don't enter into consideration when we make dining choices.

Vegans draw their line at animal products, seeing a clear ethical difference at the plant/animal divide. The most devoted among them reject use of all animal products—an extremely daunting enterprise in the technological world where ingredients of manufactured goods are hard to parse. Crayons and other art materials, film, cosmetics, tires, candles, soaps, polish, paints, glues, clothing—altogether a plethora of everyday products contain products derived from the slaughterhouse. And in farming, even gardening, it is essentially impossible to avoid killing animal life, one of the reasons I abandoned my attempt at veganism after several years. It seemed to be more posture than reality and I let it go.

Even the mystical Jains who attempt to adhere to a strict practice of nonviolence toward all living beings, and sweep the ground before them to avoid stepping on insects, can't possibly avoid eating bacteria along with their vegetarian diet. This is not to mention the soil life destroyed in even the gentlest practice of cultivating plants for food or clothing, but the Jains do their best by refusing to eat root vegetables.

Perhaps naked fruititarians come closest to achieving a vegan ideal, and that only if they pluck—but do not plant—their food source. There is a soft divide among vegans on the honey line: some reject it as an animal product, while others acknowledge that a rejection of honey would necessarily include a rejection of the pollination work of bees—which eliminates seeds, fruits and nuts from the diet as well.

Corn is a wind-pollinated crop, of course. Corn-itarianism, anyone?

Moving up the scale choices bifurcate. First there are the ovo-lacto "vegetarians," or one or the other, who include eggs and/or milk products in their diet. The self-delusion embraced by many of this bent is that their dietary choices don't include killing the source animal. This blissful ignorance nicely elides the death implicit in animal agriculture. The more practical reason for this dietary track is economic: you generate more protein per animal by eating eggs than eating hens, or utilizing milk instead of cow burger. But even this is a partial picture since something must be done with the boys, who turn up in more or less equal numbers when animals are bred. So someone eats the boy-children, or suffocates them, as is done in most chicken sorting operations.

Then, too, according to U.S. Food and Drug Administration guidelines: "An average of two rodent hairs per one hundred grams of peanut butter is allowed." There goes the vegan PB&J.

Second are the pescotarians, who include fish in their menu. This is more clearly a decision made along the consciousness scale (other than those who make the choice for presumed health reasons, since some fish are a good source of essential fatty acids). Cold blooded animals are pretty clearly less personable than warm, for lack of a better term. But here is where environment begins to play a role in our thinking. We don't easily relate to aquatic creatures because they haven't evolved in the same conditions that shaped our species.

We posit a certain amount of cuteness to vertebrate land animals that we don't accord to fish. Even watching a lizard, which exists with somewhere near the same brain function as a fish, we relate differently. We "get" how legs and arms and climbing and crawling and jumping happen much more easily than we apprehend fish movement. We connect, or think we connect, to what other terrestrials see and hear in a way that we don't feel for a fish. This figures significantly in our relationship with cetaceans. When we use the term "cold-blooded" in regard to humans it means "unfeeling" and it's easy to decide that cold-blooded creatures are just that. (In fact, etymologically, the former came from the latter assumption.) Despite the fact that whales and dolphins are warm blooded animals, their "fishiness" still affects our perception of their thinking.

Next up the scale are omnivores, of course, and the choices further divide for a multiplicity of reasons. As mentioned earlier, in our pre-civilized past there was a natural dietary spread from the Equator to the poles because vegetable foods were more plentiful in warm places and pretty scarce among the icebergs. A similar spread runs from savannah to rain forest, with more, larger animals in the plains and more tiny animals under the tree canopy. Many rain forest tribes eat more insects than vertebrates, contrary to the common assumption that rain forests are rich in higher animal life.

Today consumption of whale meat is limited to Norway, Iceland and Japan, with the afore-mentioned Makah and other indigenous groups making their limited stab at procurement.

Despite our growing understanding of the size and complexity of cetacean brains and the clear evidence of language skills, we are still hung-up on the same confusion evident in the divergent Biblical translations of "great fish" and "whale." Though it's been well known for centuries that whales are warm-blooded, air breathing mammals, there remained a kind of land/sea divide and the sense that they were really a form of fish.

One measure of cognitive function is self-awareness. A few animals seem to qualify on this score, though the dividing line is subject to debate. The most commonly used measurement device is a mirror with the test animal given a spot of color in a random location. If an elephant, for example, looks in the mirror and touches the paint spot with her trunk, it seems likely that she realizes that she's looking at her own image. (Repetition of the experiment with clear dots can establish whether the animal simply feels something odd on her skin or is actually reacting to the visual cue.)

As reported earlier, ethologists have used this and other tests to determine that elephants, great apes, bottle-nose dolphins and magpies exhibit some measure of self-awareness. The most controversial among these nominees are the dolphins because the requisite recognition-of-self reaction relies more on inference by the human observer.

Cetaceans have no means of touching themselves on a paint spot, or even of turning their heads to look, so researchers assume self awareness when the painted creature seems to notice the odd marking in a mirror and returns repeatedly to view that portion of the body, rather like a human trying on a pair of pants and turning toward a mirror to see if the cut of the cloth makes his butt look big. (Testing whales, particularly the great whales, in such fashion is more or less physically impossible.)

Another criticism of the mirror test is that it assesses perceptual rather than conceptual skills.

The step up from self-awareness is referred to as "theory of mind." This is other-awareness, the idea that another being has its own mind, its own thoughts, its own feelings. Human beings clearly embrace this concept because it is the necessary underlayment for spoken language. We understand others and seek to be understood. Even when we have difficulty comprehending what someone else thinks, we know that she does think. We might not completely connect with what another feels, but we are sure that he feels. At its core, theory of mind considers the question of whether a being can think about thought itself, entertain beliefs or entertain desires, and embrace an intentional stance—that is, the ability to interpret others' behaviors as goal-directed.

Psychologists and ethologists have set up all sorts of experiments to assess animal behaviors, often designed to add "I don't know" as an alternative to "yes" or "no" responses. (The idea being that expression of the inability to select one right answer requires self-knowledge and thinking about thinking.)

They have also searched for instances of intentional deception: one creature misdirecting another in order to gain some advantage. Such strategic behavior would be deemed evidence of thinking about the other's thinking. "I'll fool him," seems to require projection of another mind that might be fooled.

As exploration of such thinking processes goes ever deeper, scientists seem to turn up as many questions as answers. When animal A "fools" animal B, is it really thinking about the other as a thinking being, or simply provoking behavior that diverted attention. "I'll fool him" may look the same as "I'll distract him" but has a different locus. The first is in the other's mind, the second is purely an outward behavior.

Another line of thinking involves the teaching of offspring, since that appears to revolve around recognition of otherness. Here bottle-nose dolphins excel.

Mother dolphins, for example, will lead their calves to a sandy ocean bottom, poke around and roust out a flounder and bite it to slow it down so the youngster can grab the prey. They'll do this over and over again, moving in for the kill much more slowly than they would if feeding on their own. Gradually they change the lesson to where they are pointing with their noses at the likely location of a buried fish, indicating to the student where to make the grab.

While to all intents and purposes this can easily be read as teaching versus instinct, I can set against it my experience with Tuxedo Joe, a thoroughly indolent, male, neutered house cat with almost no interest in hunting. When Tux was a decade old—never a parent and never evincing an interest in rodents—we were graced with a rescued stray who promptly delivered five kittens.

When the little ones were old enough to romp some on their own, Tuxedo suddenly became a Brave Hunter and commenced to come in through the cat door with wounded mice which he would deposit in easy reach of the kittens who then played the creature to death—but mostly failed to eat. (Pomonella was always willing to help out with left-overs.) Was Tux teaching? I'd have to say, "Yes." Was it emblematic of rational thought? I don't think so. Instinct seems the more likely cause. So a dolphin teaching the kid to hunt flounder presents us with an uncertain case: it seems more deliberate and nuanced that Tuxedo's delivery of a victim, but only by degree.

Another approach to the question of cetacean intelligence and consciousness relates to culture. One of the characteristics of human society is variation in language and organizational patterns around the globe, whereas less sentient animals simply replicate behaviors.

Killer whales around Vancouver Island, BC, have been extensively studied and divided into two very distinct societies: the residents and transients. The two groups exhibit unique behaviors and despite some intermixing at the fringes, represent distinct societies.

Ongoing examination of killer whales and other cetacean species has even documented cultural drift—the tendency of populations to exhibit unique changes over time as groups disperse or intermingle.

The problem with all such studies of cetaceans, undertaken with the best of intentions and utmost scientific rigor, is their rootedness in human, terrestrial assumptions. We can't, ultimately, entirely resolve communication with any species that inhabits an entirely different reality. People like us make sense. People not like us pretty much don't.

When Europeans (*homo sapiens sapiens*) met Native Americans (also *homo sapiens sapiens*) there was no common ground on the matter of ownership of land. The very concept that anyone could own or transfer such ownership was completely off the charts for the indigenous people. Likewise, the notion that the world was a manifestation of a Great Spirit which animated everything and couldn't therefore be owned by any human was completely foreign to the European interlopers. To them, God was philosophically omnipresent, but he didn't own real estate. God was everywhere, but people owned the land.

The gulf between human and Neanderthal was presumably even wider. Paleoanthropologists devote a great deal of attention to parsing the cultural divide between those ancient branchings of our family tree. We made very different tools despite considerable overlap in territory. Though there is some evidence of our tools falling into their hands, they don't seem to have been able to copy the improvements our team demonstrated. (Barbs on arrow- and spear-heads, for example, plumb eluded them.)

They engaged in ceremonial burial of the deceased, but until 2009 paleoanthropologists believed that Neanderthals never wore jewelry—an absence interpreted to mean that they had no sense of themselves as individuals; i.e. no self-awareness. Just as we adorn ourselves today to announce personal uniqueness, it is assumed that our distant human forebears liked to play dress-up and that proto-humans who failed the fashion test lacked true consciousness. Yet recent discovery of painted shell beads associated with Neanderthal remains has cast doubt on that differentiation.

Closer to us in time, an argument has been made that what we regard as consciousness is a relatively recent phenomenon even among our species. Psychologist Julian Jaynes discussed the idea in his breakthrough work, *The Origin of Consciousness in the Breakdown of the Bicameral Mind* (Houghton Mifflin, 1990). He posits that the two hemispheres of human brains were more directly connected in an earlier time, and that they lacked self-awareness. He theorized that hallucination and visions would have been common and that those individuals would have interpreted such experiences as direct instruction from gods.

Jaynes' suggested "breakdown" began about six thousand years ago, triggered by or marked by the rise of city states and written language in Mesopotamia, and culminated perhaps three thousand years ago—about the time of the rise of the Middle Kingdom in Egypt.

(Recall my earlier discussion of mummies and the belief that the brain served no purpose—not that localization of and awareness of consciousness necessarily go hand-in hand. But try to imagine your thinking is going on somewhere else in your body. It's not easy.)

Jaynes' work has been used by many subsequent researchers exploring the nature of auditory hallucinations and several of his theories about brain structure were later confirmed with brain imaging technology. So it's entirely possible that our ancestors both heard "gods" talking and lacked what we term consciousness.

Jaynes' critics have argued that *homo sapiens sapiens* has always had consciousness, but lacked a theory of mind to explain it. That is, they were unable to think about thinking. Somehow that seems to me to beg the question. What is consciousness if it *isn't* thinking about thinking?

Another possible interpretation of Jaynes' work is that people who today believe they hear gods speaking are experiencing an ancient phenomenon involving back-and-forth communication between the left and right hemispheres of their brains.

If we aren't even clear that our own species is innately conscious how much further apart are the guiding concepts of human and whale? Reading the literature about dolphins one can get the feeling that interaction with humans amuses them. If you open yourself to the possibility, instead of assuming superiority, you can easily get the sense that the "cooperative" cetaceans in a seaquarium are having a huge amount of fun at our expense.

We're bad swimmers and don't understand the ocean or the way most of the real world works, given that the sea covers almost three quarters of our planet. We imagine that tossing a dolphin a few fish when she answers some inane request correctly is a huge incentive to their kind. We clearly have no clue about the ultimate meaning of existence.

On the other hand, of course, dolphins surely recognize that some of us are heartless and murderous. For a being which lives in a weightless infinite now, floating on a sea of sound, perhaps trusting in an infinite future on a planet without walls or wars or Walmarts, our butchery may mean very little.

It seems probable that while preconscious humans might have held beliefs, a theory of mind is required for one to consider imposing beliefs on another or to compare and decide that one's own superstitions are in some way superior. So while chimpanzees raid other clans to acquire food and mates, and ants are known to invade neighboring ant colonies, none of them attempt to impose ideology. Only humans invent belief-based rules to govern the behavior of others.

Altruism is practiced by many creatures because it is a successful survival strategy. Incest taboos have a logical basis in avoidance of genetic dead-ending. But prohibition of pleasure is the invention of a conscious, self-aware, self-righteouss, power-seeking mindset— perhaps dating back to Jaynes' breakdown when city-states and written language permitted the rise of class differences, the accumulation of wealth, the fabrication of histories and the organization of religion.

Nor is prohibition of pleasure limited to drug laws. Dancing, singing, card-playing and other forms of recreation have been subject to proscription by Christian cultures, while facial scarring, foot binding, and other imposed limitations were carried on under other belief systems elsewhere in the world. Genital mutilation of women is still widely practiced in parts of Africa, to prevent enjoyment of sexual pleasure.

We may easily have lost as much as we gained. We learned to make art and make rules and we learned to lie. Sadly, many of the lies we tell best we tell to ourselves.

fire

The windshield blasted into our faces.

We were headed for Logan International Airport in Boston to pick up the ten-year-old son of a friend in Florida. She wanted him to experience winter, and while the Manchester airport was closer, the ticket to Boston was cheaper and we were familiar with the sixty mile drive so we assented to the Beantown destination.

Susan was driving and we'd just gotten up to highway speed on old Route 101, crossing an overpass of Route 125, when the crusted ice and snow on an approaching car lifted into the air. The block was about eight inches thick and four feet square. With both cars going about 50 mph, the closing speed would have been twice that, and we had barely recognized the pending threat when it smashed through the windshield. The other driver didn't stop and was probably completely unaware of the accident.

Against all odds, Susan managed to cross the bridge and bring The Mother Van to a safe stop on the apron. Our faces and her hands on the wheel were bleeding and we were covered with shattered glass and melting ice.

"That was interesting," she said.

"Jeezus!" I responded.

After we found our eye glasses, dislodged in the collision, and determined that our wounds were super-ficial, she did a U-turn and drove slowly back home where we cleaned up and changed and took the other car, a recently acquired Dodge Dart, "Elmer," to the airport. We made it on time.

I offer that tale in part as a public service. People in snow country should be aware of the potential for that kind of accident. First, that your car might cause it (so remove the cap while you're clearing your windshield), and second, that the oncoming car might flip its lid.

But it fits into this disquisition in another way. That was our year, 1978, and our location (the 101/125 overpass) for unexpected flying objects.

One evening that spring we drove in Elmer about ten miles to Exeter to catch a film. At about 9 p.m. I was behind the wheel as we headed back out Route 101 toward Fremont. We rounded a curve, perhaps two thirds of the way home, and I noticed a brilliant light above the road ahead. I assumed it was a new, extremely bright, streetlight. The route was very familiar, so it had to be a very recent addition.

Another mile and it didn't seem much closer, but it did appear to be higher. It was an intense blue-white.

"Do you see that light?"

"Of course," Susan answered. "What is it?"

"Too high for a street light. A plane maybe?"

"Or a helicopter? It's not moving."

The light waxed as we approached. As we crossed the 125 bridge it was absolutely clear that we were seeing a very large gleaming object that hung virtually stationary above the road. I reached our usual turn, slowed, turned left on Martin Road, stopped and stepped out of the car.

"What the hell are you doing?" Susan barked. "You'll get killed."

"It would be worth it. But I'm not scared."

Directly above us, at about twice tree height (which I later measured at 70 to 80 feet—so double that) hung an object, absolutely as tangible as an airplane or dirigible. I spanned a tree that seemed to be a similar distance away with my arms and held it up for a gauge. The object was bigger, perhaps 100 to 120 feet in length, and narrower than it was long. The surface was entirely smooth and seemed to glow from within, except for a bump in the bottom surface, 10 or 15 feet across, which glowed faintly red.

It suddenly zipped off toward the south, silently disappearing from view due to the surrounding trees.

I jumped back in the car, spun around and went out on the overpass bridge which offered a clear 360-degree view of the sky. Encountering no traffic, I stopped and stepped out again. There was no bright light visible anywhere in the sky, in any direction. We waited a while, but nothing appeared.

Neither of us had a clue what we'd seen. I suggested we go home and make independent drawings and guesses at dimensions before we discussed the details, as an attempt at corroboration.

Our drawings were a match. We both estimated the length at 90 feet, which I revised upward after calculating the tree heights. We agreed on the radiant color, as well as the red protrusion, and agreed that we had seen a solid object that resembled nothing we had ever seen anywhere.

Subsequently we learned that our landlady had been interviewed by authorities a dozen years earlier, after reporting a UFO in our neighborhood. She said it had hovered above a power line and touched it with a rod-shaped extension. Her experience had been reported in *Incident at Exeter: The Story of Unidentified Objects Over America Now* (Putnam, 1966), by John G. Fuller.

I offer no explanation or theory. Thirty-two years later I don't know what to believe, except that I saw it. Someone or some "thing" had a very, very unusual flying machine. We forgot about the matter entirely until years later when I found the drawings in a file folder.

18. fuel

Over the river and through the woods.

Or—to state our trajectory more accurately—through the woods and over the river; hung a left past the bridge; a right; a left; a zig; a zag; another right; and there we were at grandma's house—just 515 miles from transom to transom. Twenty hours total driving time divided in half by a two-week visit to the home place.

That this sort of casual travel is possible is one of the miracles of 20th century technological civilization. Mobility has blessed and cursed us, enabling an endless diaspora while chaining us to our machines. In the process our dispersal may well have become the biggest psychological obstacle to creation of a sustainable society.

And while I'm pointing a finger I'm not afraid to admit my own guilt. From 1976 to 2000 I spent a great deal of time on the road, for pleasure. Cheap oil let me visit the Grand Canyon and the Badlands, New Mexican mesas and Aztec ruins, Big Sur and the Olympic Peninsula, Vancouver and Fairbanks and Anchorage, the Yukon, New Orleans and Chicago, San Franciso and Washington, DC, Tijuana and Newfoundland, the Little Big Horn and the Saw Tooth range, the Okefenokee and the Louisiana bayou. I've canoed in every Great Lake and many of North America's river systems. I've hiked in Yellowstone and the Snake River Canyon and the Chiricahuas and the White Mountains and the Green

Mountains and the Cascades, the Catskills and the Sierra Nevada and the Sand Hills and the Ozarks, of course the Southern Applachians where I live, and too many more places to mention. It was a grand adventure and it was cheap.

Taking the long view—disastrously cheap.

In addition to Susan's inclination toward travel and my own willingness, there's a longer-term picture to consider as well. My parents met and married in Florida, though Dad was born in Chicago (as was I). The introduction occurred because my Mom had worked in New York City for a couple of years, where she met Dad's cousin (also from the Chicago area) who suggested the two get together after Mom returned to her high school home town, Orlando, where Dad was building homes and breeding Shetland sheep dogs. Mom was born and halfway raised in Pittsburgh. I had moved to New Hampshire and then North Carolina with a year-long stop-over in Arizona. But my trajectory had included junior high in Long Island, New York, and high school in Florida, with a couple of years of college in Atlanta.

Susan's parents settled on her grandmother's farm-turned-suburbia in Ohio, and, as previously mentioned, aunts, uncles, cousins and two siblings stayed near, but her other brother moved to Tucson, then Portland, Oregon. A niece and nephews spun out to Washington, DC, Knoxville and Salt Lake City/Dallas/ Atlanta, respectively. Neither of our families was particularly atypical for the post-WWII years. We spread out and dissolved the extended families of past generations. We did so because we could—often for better job prospects, sometimes on whims, for love, or, pretty often, simply to shake off the past.

Cheap energy made cost no real object, and that same cheap energy made family visits, shared holidays, weddings and funerals and graduations and other base-touching reasonably affordable.

But the families were fragmented despite phone calls and (increasingly rare) written letters. (E-mail has lately abetted better and more frequent contact for many.) The easy distancing could engender real difficulty when a physically remote mother or father needed nursing care and the lack of nearby grandparents shifted more children into daycare.

Whether this social fracturing has been, on the whole, good or bad is open to debate, but the fact that cheap oil had social consequences is not.

That holiday journey traversed a landscape in transition. Farmland was sprouting subdivisions as thick and fast as springtime weeds, particularly along the Interstate arteries. The previous week, one of Susan's brothers went to an auction of the farm which he (and their father before him) worked on as a young man. The gavel came down to the tune of one and a half million bucks, paid by a developer hell-bent on suburbia.

We have painted ourselves into a very difficult corner as cities metastasize into surrounding healthy tissue. The sprawl enabled by fossil fuel combustion has built us into a dependence on that technology that becomes harder and harder to break.

Look at the conundrum: Cheap mobility facilitates both commuting and distribution of goods. Easy commuting drives up the use-value of land far outside the cities, a change which also raises property taxes. At the same time, the distribution network permits import of food from lower valued land (usually with lower priced labor). Beleaguered farmers facing underpriced competition and overpriced land are understandably tempted to liquidate. The whole scheme floats on cut-rate oil.

Each new home on former farm land further entrenches political support for the status quo. People who have invested their savings in a home and who are dependent on a distant job to keep up mortgage payments are vested in the present cheap-oil economy.

Adding insult to the internal combustion injury of the biosphere, the average size of new homes in the U.S. is growing. More heated space will require more heat for decades into the future. Even construction methods are affected, as when cost/benefit considerations dictate the return on insulation or insulating windows vis-a-vis cheap energy.

At the same time, inexpensive oil encourages investment in inefficient vehicles, and—via conversion to electricity—in inefficient appliances of all sorts. Each consumer decision against conservation results in further stasis. A new auto which uses twice as much fuel as an alternative model locks that demand into our energy equation for twenty years or more. Ditto for refrigerators, freezers, ranges, water heaters and a host of smaller gadgets.

Meanwhile the supply lines for food grow ever longer, and more and more of the fertilizer supply comes directly from what Thom Hartmann aptly referred to as "ancient sunlight"—fossil fuels stored up over millennia.

The U.S. has created one of the least efficient technological societies on the planet. While other countries have pushed fuel prices up through taxation to encourage conservation, we have intentionally kept fuel prices low—an intentional subsidy to drivers, industry and agriculture.

Moreover, we subsidize oil with tax money for military intervention as well as funding health care costs incurred by pollution victims. Our inefficient vehicles and high reliance on automobile use has created a childhood asthma epidemic, directly attributable to auto exhaust.

More subtly, oil costs are externalized in the form of forest and agricultural decline resulting from acid deposition, nitrous oxides and low-level ozone. As I write this sentence on January 15, 2010, citizens of Asheville are being warned not to engage in overmuch outdoor activity because we are Code Yellow.

This in a city once famous for it's healthful air.

In order to move toward true sustainability we must—by definition—decouple our lives from dependence on non-renewable resources, but the political will for such sweeping change is conspicuously rare.

Though I treasure the chance to spend some holiday time with distant family members and friends, it is impossible to shake a sense of foreboding. The policies that made that visit possible will make our entire economic structure impossible in the not-too-distant future as the oil runs out. We are building toward a crash of monumental proportions, on a scale that could easily dwarf the experience of the Great Depression or the current Great Recession. At least in the 1930s most of us lived closer to the farm, to our work, to our families.

We chose to believe that low fuel prices were a social good. Our elected officials made sure that continued, at least in part because it was an easy issue. People notice how much it costs to fill their tanks and fill their grocery bags and if those prices jump up an opposition candidate can promise to knock things back down. So incumbents keep the lid on. The larger costs are hidden and spread out as hospital bills, acidified lakes, military intervention, and more—invisible in plain sight because they are diffuse and the dots haven't been well connected in the public mind.

Scattered to the wind, we are, and living amidst strangers. We have been fooled by cheap energy into choices we might soon regret. I fear it may be a very, very long journey home.

19. grass

Libertarians are Republicans who smoke dope.

Modern Republicans tend to be people who believe in legislating morality. They're prone to be law-n-order sorts who believe in strict behavioral codes and believe that the law can enforce that code—particularly for the untrust-worthy hoi polloi. Hence their historic support for school prayer, an abortion ban, exclusion of homosexuals from the military, and much more.

That's why recreational drug use is likely to lead them to a Libertarian stance. If something isn't part of their moral code, it shouldn't be enshrined in law. This tendency toward Libertarianism is greatly abetted by their relationship with business regulation, which assumes good behavior on the part of management in response to the perceived best interest of stock holders. Generally speaking, Republicans tend to distrust government, and see it as too likely to be controlled by "special interests."

Modern Democrats, on the other hand, tend to believe in legislation that leaves moral choices up to the individual, only imposing strict rules where such choices infringe on others. Democrats share an inclination to keep religion out of public spaces (leaving maximum freedom to each person to pray or not), to make abortion a private choice, to not ask about sexual orientation and so forth. Their anti-Libertarianism is driven by the question of infringement on others and a perceived need to protect the commons from the actions of individuals. Hence, also, Democrats share an inclination to regulate business which is seen as likely to exploit workers and cheat customers in the name of profit. Generally speaking, they tend to trust government to work for the many, casting "special interests" as a collective effort which ends up supporting the general good.

Libertarian-leaning Dems are apt to become Greens, seeking strict regulation of commercial polluters but a hands-off policy on personal behavior.

Of course, there are a thousand variations, and like any generalization, mine can be shot full of holes. But I want to use the above for a starting point on why drug laws have had such a devastating effect on our society.

Many members of my generation of Americans smoked pot. According to recent surveys, we still do. Possibly more than half are still lighting up, and a clear majority are entirely willing to do so in the right circumstances: i.e. getting caught can be a big hassle, but absent enforcement or in trusted company, why not? The criminalization of marijuana which began early in the 20th century has completely failed to stem the popularity of the drug.

In the 1970s I didn't know anyone about whom I could say with certainty that they had not ever tried marijuana, except my Mom. Probably.

My Dad certainly had, while he was a member of the Florida House of Representatives from 1968 to 1970. He obtained the drug from a friendly Florida State University student, rented a motel room in Tallahassee and proceeded to get stoned. Ever afterward he was a mild advocate of decriminalization, saying that he saw nothing much to recommend it, but certainly no reason for a ban.

One of Dad's heroes, the conservative ideologue William F. Buckley, Jr., did essentially the same thing and arrived at the same conclusion, only with a bigger budget. Buckley sailed his yacht out beyond territorial waters and reportedly obtained the grass on the high seas so that his experimentation was completely legal. It seems like a lot of trouble to do something so simple, but then Buckley was planning to write about it, whereas Dad kept the smoking experiment pretty much to himself.

Buckley then wrote, ""Marijuana never kicks down your door in the middle of the night. Marijuana never locks up sick and dying people, does not suppress medical research, does not peek in bedroom windows. Even if one takes every reefer madness allegation of the prohibitionists at face value, marijuana prohibition has done far more harm to far more people than marijuana ever could."

Moreover, a high percentage of people I knew in the 1970s grew pot, or made an effort in that direction. While organic farmers who were tuned in to heirloom varieties passed along seeds for "mortgage lifter" tomatoes, there was little doubt among many of my peers that there were easier ways to lift a mortgage.

In New England in the 1970s and 1980s there was a well organized growers network through which rural land owners produced crops on contract with distributors. The cleverest of the growers often cultivated their cash crop on vacant land adjacent to their own acreage to avoid the possibility of being charged with possession of felonious quantities of weed. I can only assume that such networks were common, nationwide.

A few friends ran afoul of the law, of course, and had to extricate themselves at great expense in time and money. As had always been true and remains true today, those with money generally walked, particularly if they were white. The poor and the non-white go to prison.

While marijuana's physical effect on most people is decidedly benign, widespread indulgence in illegal drug use has had a very toxic effect on the body politic. Pot laws and the rhetoric supporting those strictures created a major disconnect between reality and the law. Authorities who claimed pot was addictive or a gateway drug to harder stuff were proved wrong over and over again. We all knew we were being lied to, but we also knew we were choosing to become criminals. We began to fear the police and question authority more broadly.

A scofflaw attitude leaks into other areas of life. If a person believes that one law doesn't apply to him, there's a distinct inclination to pick and choose among other laws. This is the basis for Lawrence Lessig's arguments for rethinking of copyright laws in light of the rise of music file sharing among today's youth. Sharing music is even easier than growing pot, and we are tempting a generation into a criminal mindset.

It recalls to mind an aphorism familiar to those in the legal profession, that "hard cases make bad law." This comes from the appellate courts where it is recognized that when judges are forced to stretch for a just solution in a particularly difficult decision, the unintended consequence may be to open a path to future problems. The opposite is also true, I think. Bad law turns many people into hard cases. And our drug laws are very bad.

Participation in democracy diminished after the 1960s and didn't return to the pre-60s high until 2008. The disaffection certainly had many causes including widespread disillusionment following the Kennedy assassinations and probably the rise of TV viewership (a.k.a. the "plug-in drug.") But the drug laws alienated a wide swath of people with no good effect. As a result of the disaffection we were easily prey to the feel-good nostrums of Ronald Reagan who taught that government was the problem, rather than an historically proven source of solutions.

The Bush wars, the federal failure following Katrina, and the Great Recession that have bankrupted our nation are a direct result of citizen disengagement and a broad distrust of government. The War on Drugs has been one of the biggest policy failures in the history of the United States.

For more than a century this country has attempted to enforce prohibition based on what amounts to a religious stricture. Hedonism runs against traditional Judeo-Christian notions of propriety. People, particularly the common folk, shouldn't be allowed to indulge their desire to have unauthorized fun.

It is the same religious asceticism that outlawed alcoholic beverage sale, manufacture, and transportation from 1919 to 1933, as mandated in the Eighteenth Amendment to the United States Constitution. Of course, prohibition didn't end alcohol consumption, it just drove it underground and rendered it considerably more dangerous. Unscrupulous vendors were not particularly worried about the toxicity of their product. Dodging federal agents was a much higher priority than quality control.

That era saw the rise of gang activity, gun violence, police corruption, smuggling, arrests, adjudication and imprisonment—all of which are immediately familiar to the modern reader. Prohibition was ever thus.

Unlike the ban on other substances, initiated at the federal level during the same era, alcohol prohibition was opposed by powerful financial (and therefore political) interests. Then, after alcohol prohibition was overturned, distillers had every reason to support a continuing ban on competing recreational drugs. As the pharmaceutical companies grew in power, they too had a strong incentive to prevent self-help on the part of potential prescription customers. (Note that Coca Cola® included cocaine as an ingredient in the early 20th century—a competing drug delivery system soon eliminated by federal authorities.)

At the same time the threat posed by hemp production to forestry and chemical companies also required suppression of marijuana cultivation. Hemp fiber is superior to wood fiber for paper making and better than cotton for durable clothing. Hemp oil can replace many of the products formerly crafted from whale blubber and, later, petroleum. The financial cards were thus heavily stacked toward suppression of hemp culture, despite a preexisting federal boosterism for the growing of hemp for naval (and other military) rope.

In recent decades the widespread privatization of our prison system has created another huge lobby for continuation of prohibition. We enjoy the highest per capita incarceration rate in the world, which not only funds those prisons directly, but provides a low cost work force for the corporations which use prison labor in order to reduce their employment costs—a sweet deal all the way around.

Low-level participants in the drug supply network are thus reduced to permanent unemployment, since a felony conviction makes finding a real job approximately impossible and making a return to drug dealing a highly worthwhile gamble. We also punish their families and help create a subsequent generation which lives in poverty and sees drug dealing as a short-cut to some measure of success.

But ethical preferences are malleable and when we discuss the war on drugs it's helpful to bear in mind the longer history of global drug policy. Back in the days when the British Empire ruled the seas, trade with China became highly problematic. There was a burgeoning Western demand for Oriental spices, tea and silk, but the Brits had very little to offer a civilization that had been happily self-sufficient for thousands of years. Trade requires mutuality of desire.

How to woo Wu? Say it with flowers, of course.

In the 1800s the clever Brits (with American collaboration) injected consumer demand into the Chinese market in the form of opium. With its solid foothold on the Indian subcontinent, Great Britain was gatekeeper for the world's opiates. Opium and its derivatives (codeine, morphine and heroin) represent the ideal commodity: a consumer product stripped bare of aesthetics, fad or advertising spin. Adam Smith's invisible hand packs a pipe, plops a few drops on the tongue, squeezes a syringe—and reaches out for more. Trade blossomed.

China's leaders balked. Becoming a nation of junkies was a higher price than they were ready to pay. The spurned suitor declared war, and before long Great Britain controlled China's coast and imposed its drug dealing on that reluctant polity. Hong Kong, given its walking papers in the waning light of the 1990s, was the last official relic of the Opium Wars. That very successful War *for* Drugs was finally over.

How things change! In the 20th century U.S. adopted the role played by China in that earlier era and so we have had the War *on* Drugs. (A man points to a skillet on a stove and says, "This is drugs." He cracks an egg and dumps the yolk into the hot skillet. As the egg begins to fry and sizzle, he concludes, "This is your brain on drugs. Any questions?") And oh, how things stay the same! The international drug trade remains a wonderful excuse for military intervention.

Now as then, the Obama administration is ramping up a war in Afghanistan against forces substantially funded by the heroin trade. In effect we are being taxed twice: once at the federal level to support the war, and again at the state and local level to suppress heroin consumption that is funding the other side in that same war. Our war on drugs helpfully drives up the prices and therefore the profits of those we oppose in our war on terrorism. While Obama's drug czar has indicated that it is the administration's goal to end the war on drugs, to date there has been little change from the policies of their predecessors.

George W. Bush who allegedly snorted, toked, drank to excess and evaded drug testing while in the Air National Guard, engaged in the same suppression efforts as his predecessor administrations. The chief effect seems to have been strengthening the drug cartels which has splintered Mexican politics and collaterally greatly increased demand for smuggled American weapons.

The ("didn't inhale") Clinton administration, egged on by Drug Czar Gen. Barry McCaffrey, pumped a couple of billion dollars into warfare in Colombia. The stated goal was stemming the flow of drugs into American noses, mouths and veins. The unstated goal appeared to be saving Colombia's central government, which had ceded control of nearly half the country to rebel forces.

Does all of this reflect a drug policy or a covert military campaign? Do highly successful rebels constitute a drug army or a popular insurgency? Could it be that the Afghan al Qaeda—like the Colombian revolutionaries and the Kosovar Liberation Army before them—are selling drugs to buy guns, rather than arming themselves to protect a drug concession? What of the Mexican cartels who may or may not control wide swaths of the national army and police forces?

And where, exactly, do American interests fit in here? Post-war reports from Kosovo suggested that our policy there included tacit acceptance of a continuing heroin trade. This burgeoning business not only funded the Kosovo Liberation Army but also appears to have underwritten the uprising in Chechnya. Is there some magic formula that explains why repressed Eastern Europeans are freedom fighters, whereas repressed South

American peasants are only trying to get Johnny hooked on hard stuff? Are we sanctioning, or ignoring, or attempting to curtail opium production in Afghanistan? Or all three? Did we ignore the heroin trade when it was financing the mujahedeen's war against the Soviet Union? Is there some reason why the flow of heroin into Europe or out of the mid-east is geopolitically correct, while the flow of cocaine into the U.S. is abhorrent?

One real effect of the ongoing War on Drugs is certain: Street prices will continue to rise.

And here the Opium Wars hit home. Low level property crimes in the U.S. without much exception involve drugs. People steal to support drug habits that would be easily affordable absent prohibition. Gangs proliferate. How many gangs have formed around delivery of cigarettes? Murders will occur. When was the last time you heard about a shoot-out concerning delivery of booze? It doesn't matter what moralizing we bring to bear on the question of drug use itself. The inescapable truth is that if other recreational drugs were regulated like alcohol and tobacco are controlled, the cost to society would absolutely, inarguably be much lower.

It took a little more than a decade for Americans to decide that they had seen enough body bags, enough innocent blood shed, to demand an end to the Vietnam War. The Wall in Washington offers mute testament to our slow reflexes and wasted lives. The Iraq/Afghan incursion will reach its decade soon, with no better show of success and the voices calling for an end to the madness are gaining strength. The War on Drugs has been waged for a century now, and still the killing rages, the collateral damage grows, and the stated goals remain elusive as ever.

Maybe religion is not a good reason to create prohibitions against personal behavior?

Portugal decriminalized drug use in 2001, and follow-up studies have shown that drug use has decreased. A paper, published by the Cato Institute in April, 2009, found that in the five years after personal possession was decriminalized, illegal drug use among teens in Portugal declined as did the rate of new HIV infections caused by sharing of dirty needles. The number of people asking for drug addiction treatment more than doubled.

Glenn Greenwald, an attorney and author who is fluent in Portuguese conducted the research and wrote the paper. He said, "Judging by every metric, decriminalization in Portugal has been a resounding success. It has enabled the Portuguese government to manage and control the drug problem far better than virtually every other Western country does."

Still, we persist because Republicans want to control your behavior and Democrats are afraid to appear "soft" on crime.

20. song

Humpbacks sing across the seven seas.

Nor do they sing alone. Blue, fin and minke whale songs have also been shown to travel thousands of miles. In the days before human technologies introduced sonic pollution in the form of ship engine, sonar and drilling noise, whales near Puerto Rico could rhapsodize to others off Newfoundland or Spain and hear melodies in return.

Dr. Christopher Clark of Cornell University has been listening to whales, hoping to decipher their enigmatic songs. Using hydrophones he has collected thousands of acoustical tracks and he reports, "The range is enormous. They have voices that span an entire ocean."

Humans are uncertain whether whales thousands of kilometers apart communicate directly with each other, or what their messages contain. But it is well understood by researchers like Clark that a whale's consciousness and sense of self is based on sound, not sight.

Whale sound is loud and low—often so low that they are totally inaudible to human ears—and perfectly adapted for long distance travel in water and such a pan-oceanic range works well for creatures that rely on reflected sound, rather than light, to navigate, to seek food and to find mates.

The best known and most studied of whale vocalizations are those of the male Humpback, whose songs are limited to the breeding season, with the obvious implication that they have some importance in mate selection. Other sounds seem less specific, and scientists have observed that the question of whales' enjoyment of their singing is an untestable hypothesis.

Male blue whales, possessed of the earth's loudest and deepest voices, also sing during mating season and their songs have deepened in recent decades. Studies indicate that their song pitch has dropped by one third since 1960. Because this change has tracked the resurgence of their numbers in our post-whaling era, and because deep-pitched sounds don't travel as far, some scientists speculate that reduced need for long distance communication has permitted them to return to ancient frequencies.

Other researchers note that it could simply be cultural, since it is well known that humpbacks learn songs from one another. Perhaps deep voiced singers are in fashion. Some hypothesize that because larger, ostensibly more virile whales tend to produce deeper songs, other males may be trying to emulate them, just as human males might lower their voices when trying to impress a woman. "The exciting possibility, I think, is that they're all listening to each other," observed Hal Whitehead, a Dalhousie University biologist who specializes in cetacean communication. "This is a worldwide cultural phenomenon, and that's very cool."

The latest research indicates that humpback songs contain elements of human language. While researchers will not yet say that whales have their own language *per se*, they have noted striking similarities between human language and the songs of humpback whales. Both humans and whales communicate by using discrete sound units that are arranged within a hierarchical structure. That is to say, a document is composed of paragraphs consisting of sentences, which comprise clauses created with words.

Humpback songs comprise themes of assembled phrases, with each phrase consisting of units. In sum this is very similar to syntax, the grammatical arrangement of words within sentences.

Computer analysis and human observers agree that whale songs are not only hierarchical, but they convey around one bit of information per second. Humans generate ten bits of information, or variance, for every word that is spoken at a rate of something like one to two words per second. As with human speech, it can be inferred that each whale vowel or syllable contains at least a few bits of data that correspond to different sounds produced by the speaker.

One versus ten may seem to be extremely slow, but whales communicate in water over vast distances. Water carries sound four times faster than air, while for all we might infer, time itself is different for cetacean species.

Jennifer Miksis-Olds, a research associate at the University of Massachusetts at Dartmouth's School for Marine Science and Technology, is one of the few people in the world who have used information theory to study whale songs. Miksis-Olds said that marine mammal songs and sounds cannot be classified at present as language, but admits that scientists do not yet understand the meaning of whale songs and that more research is needed. "All animals in the same population seem to share the same song type ... However, the details of this are unknown. What the songs mean is, again, unknown."

Complications in the possible interpretation of another species' language, if "language" it is, are many and profound. Linguists have noted, for example, that our own speech consists of a flow of sounds not easily dissevered into discrete words. The listener is normally able to parse meaning from a familiar language, but a recording is not easily sliced into distinct words. They all flow together. The spaces which frame written words are not present in spoken language.

On the other hand, linguists, and even lay listeners, can interpret unfamiliar human languages as meaningful, simply due to the predictable cadence of human vocalizations. Human speech sounds like human speech and we can guess that there is meaning in the flow, even if we are ignorant of the specifics.

With whales we are dealing with alien intelligence, vocalizing in a very different medium, discussing (if they are) an entirely different world-view based on utterly different physical parameters. It seems to me that there should be no question about their consciousness or their intellectual acumen, but I can offer no proof.

Another creature offers an interesting counter-point to my assertions of whale consciousness. If you are lucky enough to live where katydids sing a summer serenade, I wonder if you have taken time to listen to their plainsong. Musical magic echoes through the treetops in late summer, a song sixty feet deep and a thousand miles long. A song like a river. A song like a storm system swirling through hot August nights.

Katydids are leaf-green kin of grasshoppers and crickets, and like those less-eloquent cousins, they generate sound by rubbing their legs on their wings. But whereas grasshoppers make intermittent sounds, and crickets hiccup their way through a slow, "Dueling Banjos" sort of twitter, katydids are nothing short of symphonic.

When katydids mature in mid to late August, their moonlight sonata stretches from Georgia to New England. As with whales, most human ears don't sense much nuance in their calls, which someone a long time ago anthropomorphized into "Katy-did, Katy-did." A careful listener will hear more than that, however. The sound is a twittering buzz. It conveys the ping of a finger rippling tines on a stiff comb, the trill of fine bubbles in an aquarium, and something of the sibilant hiss emitted by snakes or pressure cookers.

If you climb to a promontory overlooking a mountain cove, you can apprehend the symphony in all its ululating grandeur. Swells of sound exactly like ocean waves roll through the forest canopy. The volume builds on a distant ridge and advances through the bowl of the valley, crashing around you and then continuing on.

Immersed in such wild harmony, I envision the waves beginning with a sort of jazz solo: One katydid, singing along like all the others, is suddenly moved to blast a few jubilant notes, and his immediate neighbors join in. The joyful crescendo is picked up by others, quickly building to the wave of sound that courses through the trees.

How far does that message travel? I guess it might resonate the length of the mountain chain, to the end of katydid-land, and then bounce like a boat wake off the lake shore. Think of it. One katydid in a dogwood in northeast Georgia playing a solo that reverberates in Lake Placid, New York. Another in a sugar maple overlooking Walden Pond beating out a rhythm that echoes all the way down to Roanoke, Blowing Rock, Cherokee and Chattanooga.

What are we to make of it? What do katydids make of the vibrations from distant mountains that ripple through the night? Maybe it's the simple message that we are not alone. Others share in katydidness and/or humanity. Others share the dance of life, sing out in joy or suffer hardship, and find consolation in the blues.

When, someday, we hear and understand a coherent message from the stars, as one must suppose we eventually will, I wonder if it, too, will be music. I wonder if alien bodies boogie down in the light of distant suns, beat drums, blow horns and pour out their passions in song. I wonder if our answer will be a tedious political speech—or a global burst of drumming. I wonder if aliens fifty light years distant are already tuned in to Glenn

Miller and Benny Goodman, while others, nearer in time and space, pick up on MTV. And I wonder if we are already hearing that sort of coherent message from the cetaceans on earth, only we are too self-absorbed to comprehend what we are hearing.

I look at a katydid, its big-eyed green face as weirdly different from my own as any flying-saucer pilot's could ever be; its six legs and wings, wiggly abdomen and curved antennae unlike my own body in almost every detail. And yet I hear their music, and I get it. The world is vibrantly alive and singing its heart out.

"How can I keep from singing?"

ice

Edges are where change occurs.

We were down below the cliffs, jumping from rock to rock, watching gulls and gazing south along the Newfoundland coast toward the Tablelands, where stark, rusted, barren mountains tumble into the sea. Those mountains tell a long story about tectonic movement and the collision of continents over billions of years. They are naturally treeless because their bedrock is infertile, even slightly toxic: material which was shoved up from the earth's core when North America and Europe did the Bump in a distant age.

Fascinating stuff, this floating crust we call terra firma. It comes up from the depths as lava or is lifted by forces so powerful that solid stone acts like warm plastic: bending, folding, mutilating in rifts and layers and incursions. Elsewhere it is falling back into the middle, drawn down into oceanic trenches as continents spread apart. The cliff wall behind me exhibits warped layers of slate interspersed with crystal quartz and greenish malachite intrusions. No new rock here, just movement.

A little more accessible, in time scale at least, are the boulders recently sheared from the cliff. One must exercise caution walking below such escarpments — new material chunks off with each retreat of the sea. Between the rocks, ephemeral tide pools form a slow slide show, changing a little with every flow and ebb. Tiny shrimp, crabs, mussels and an occasional fish wiggle out an existence.

Across the inlet I spot a moored boat canted slightly to one side, grounded until the incoming water floats it once again. The sight is familiar in Maritime Canada as it is in any tidal basin or port around the world. A modest vessel needn't be moored in a deep channel if it is designed to sit comfortably on the strand.

Back in 1962, President John F. Kennedy famously remarked that, "A rising tide lifts all boats." Not a surprising observation for a sailor to make, but it caught on, hopefully assumed to be an apt metaphor for a growing economy. However, Mr. Kennedy was wrong. That was very clear to me as I slid around on the seaweedy sea bed considering plate tectonics and barnacles, treeless slopes and fish.

The truth is that a rising tide lifts all boats *here*. In order for the tide to rise in Rocky Harbor, it has to fall somewhere else. No new water, just movement. That is where Kennedy's metaphor fails, and why it continues to misinform current economic debate. When used as an argument that worldwide economic growth can lift the developing world out of poverty without sacrifice on the part of the rich, it embraces the all-boats fallacy. It conveniently begs the question of how the tide can be everywhere high at once.

Economic growth does not occur in a vacuum. The Western technocracies are rich because they appropriate 75 percent of the world's wealth, all of which is based on natural resources. The tide is high here because we make it low somewhere else.

Current, carefully reasoned scientific guesses suggest we now exceed the earth's natural resource production and carbon dioxide absorption limits by at least 40 percent. That is, we are living on our capital, not interest from our ecologic accounts. Western resource use alone already outpaces sustainable levels, even if the poorest two thirds of the world's people simply disappeared. (We're not lifting *their* boats, we're punching holes in them.)

Efficiency gains are often touted as a way around such zero-sum accounting, but while potentially helpful, they are not in and of themselves a solution. The average auto in the U.S. is more efficient than thirty years ago (despite the popularity of gas-guzzling SUVs), but we drive many more miles and consequently use more fuel. Any real progress toward resource equity will require reduced demand from the rich.

A few hours later, the vessel across the harbor swings on it's mooring line as a freshening Westerly blows waves into a white froth against the black slate palisades. All the boats in Rocky Harbor, Newfoundland are on the rise, while at Dover beach "where ignorant armies clash by night," perched on distant England's white-cliffed shore, the tide is running out.

21. edges

Susan wanted to visit the Vikings.

In the late summer of 1999 she decided we'd travel to the only confirmed Viking settlement in North America, a venture that involved a long ferry ride from Portland, Maine, to Yarmouth, Nova Scotia, camping our way 435 miles east to Cape Breton Island, thence a second ferry from Sydney to Port aux Basques, Newfoundland, and journeying north 425 miles to L'Anse aux Meadows. Nearly everywhere we drove or hiked we saw moose, formerly imported as target animals and now well established in the absence of predators other than human hunters.

The coastline of Newfoundland is spectacularly rugged. Surging waves send plumes of spray, sometimes including small boulders, up over the rocky benches that constituted much of the shoreline. At the north-most point we could see icebergs drifting off the coast of Labrador and a pod of orcas crested in the steel-blue sea, clearly visible only through binoculars.

According to Norse sagas, in A.D. 1000 (also known as Y1K for those who are keeping score at home), Leif Eriksson, eldest son of the notorious Erik the Red, sailed from Greenland for lands previously sighted by Bjarni Herjolfsson. (Leif's name was easier to pronounce, so he got credit for the discovery.) Leif and his crew landed on the shores of a beautiful island he named Vinland where a crew member (referred to in the aforementioned sagas as Tyrker the German) discovered plentiful grapes. Later Viking voyagers to Vinland met natives whom they called skraeling, which has always sounded like a baitfish name to me but meant something like "undocumented worker"in Norse.

Ever since the Viking sagas became widely known in the 19th century, scholars have debated their veracity while theorists have named locations from Labrador to Florida as Tyrker's original Vinland. Stones incised with Norse runes turned up in Maine and putative Viking artifacts have even been identified in the midwest— perhaps traded from tribe to tribe amongst the mischievous inland skraelings to confuse future archaeologists. Or perhaps, like dinosaur bones, they were placed on earth by God to test our faith.

Then, in 1960, undeniable proof of Vikings in North America came to light at L'Anse aux Meadows. Numerous Norse Viking hardware items and irrefutably Icelandic-style house foundations offered proof that Vikings had landed, and briefly settled, in North America at least 500 years before Columbus stumbled into Hispañola. Viking Day parades have yet to catch on, however.

Continuing archaeological work has revealed more than 300 years of sporadic contact between the Greenlandic Norse and various Indian, Inuit, and other Native American skraelings, concentrated primarily in the Canadian Arctic, but with tantalizing traces in South America as well. Professor Helmut Zettl of Vienna, Austria, has theorized that Vikings may have been the progenitors of the Incas, who emerged at about the same time as Leif Erikson's voyages, to become the rulers of a large empire in South America. In the village of Chirimoto in the Chachapoyas region of Peru's Amazonas department, Zettl discovered a population, half of which were blue-eyed and blond He has reported that they speak Quechua, the Inca language, with a sprinkling of Norse words. The same claim is made for Garrison Keillor's hometown of Lake Woebegone, Minnesota, with its Norwegian batchelor farmers. Myth mingles with reality in curious ways.

Being seafarers and heavily dependent on ocean products for survival in the short growing season and harsh environs of Iceland, Greenland and Norway, descendants of the Vikings took readily to whaling.

Located as they are in the summer feeding ground of several species of cetaceans, those nations became leaders in the modern whaling industry, slaughtering hundreds of thousands of the giants up until an international whaling moratorium was declared in 1966, followed by a ban imposed in 1985. By this point northern right whales had all but disappeared from Icelandic waters, hunted to near extinction.

In 2006 Iceland decided to thumb its nose at the International Whaling Commission and resumed commercial butchery, albeit on a more limited scale than in the 20th century. The nation was severely criticized for permitting the slaughter of animals listed as threatened under international treaty, but the government simply shrugged.

There was a bit of good news in the spring of 2009 when research was presented at the Acoustical Society of America, that revealed the distinctive calls of North Atlantic right whales had been detected in a former whaling ground off the southeastern tip of Greenland.

Right whales in the area were presumed extinct due to hunting as far back as the late 19th century, and in the past 50 years, only two right whales had been spotted in the area.

Following our visit to the northern tip of Newfoundland, we headed back south, stopped at Gros

Morne National Park and spent some days idling and hiking. One morning we set off from Long Pond on the Green Gardens Trail, which took us across tablelands (flat, as the name suggests) to a coastal area of volcanic rocks and sheer cliffs.

For whatever reason, Susan had really gotten on my case the previous night and kept it up as we hiked that day. She belittled my writer/editor career and revisited one of her frequent themes: Warren Wilson College, where I had founded an environmental journal and was writing copy for two radio programs, was a third-rate, two-bit school and my employment there was entirely predictable, since I was a third-rate, two-bit writer. I needed to get real, recognize that I would never amount to anything as an author, and refocus on construction work. It's difficult to convey the particular way she was able to insert her judgmental knife and twist it to achieve maximum pain, but her sharp-tongued criticism had become increasingly difficult to tolerate.

What I understand today is that my failure to assert my feelings, even to identify them, had bottled up a world of hurt. My mother had quietly martyred herself to marriage, modeling a placidity in the face of my father's verbal abuse, and teaching me by example. I yearned for approval from Dad that I never received and sucked up to his criticism, visibly dutiful and yet privately rebellious. I dropped out of the life he'd envisioned, his chase for money and influence, but still hungered for his love. The hunger had continued to play out in my relationship with Susan, all deep beneath my awareness.

In that moment though, I felt very deeply hurt, defensive, unfairly criticized and, finally, profoundly angry. It was one of those "scales fall from eyes" moments when the world suddenly looks very different. As we hiked along a seaside cliff, miles from civilization, miles from any other observable human beings, I entertained an extremely disquieting idea.

It suddenly struck me that I could very easily push Susan off one of the sheer cliffs we were traversing, whence she would plunge to her death, and that no one would ever be the wiser. I was instantly repelled by the idea and felt certain I could never live with myself if I actually followed through.

Even as I weighed that unthinkable thought, I knew I really needed to get out of a relationship that was pushing me to such an extreme. From that hour forward, I was mentally headed out the door.

It was easier said than done, however. I was afraid to tell Susan that I wanted out—fearful of her rage, scared to assert my preferences after twenty-three years in her thrall. I gradually came to understand how completely I had given over control of my life to her.

Back in North Carolina I continued to contemplate escape and finally, in mid-December, gritted my teeth and wrote a three page letter declaring that it was my intention to separate our lives. But it was constructed as a missive of apology. "I have tried my best, for years now, and it is entirely clear that I am never going to make you happy," I wrote, filling in salient details of recent failures in my behavior. I concluded by stating that I believed it was best that we part and that I would make arrangements for my move as quickly as possible. Looking back I understand that I was still not able to express my own hurt and anger and had to couch my "Dear John" by accepting the entire blame for our problems.

When I left for the college that morning Susan was already at work in her clay studio, so I put the letter in a recycled business envelope, wrote her name with felt-tip marker and perched it on the kitchen range, directly in her path from the front door to the pissoir where I knew she couldn't miss it.

When I returned home that evening the letter was where I had left it and given that I'd put it in a recycled, sliced-open envelope, there was no way to know if she'd seen it or read it. I chickened out. I picked it up and carried it up to my room, shredded it and put it in the recycle bin. It wasn't until three months later that I figured out that she had almost certainly read it and replaced it on the stove. In any event, she didn't mention it and I had withdrawn into my shell again, once more afraid to broach the subject while our relationship seemed to deteriorate by the day.

Before our journey to Newfoundland, Susan had been troubled by a lump and discomfort in her right breast. She'd visited her doctor and submitted to a mammogram which was diagnosed as normal. Eighteen years earlier she'd had a bout with some uncomfortable mammary cysts and her doctor told her it was probably a recurrence of the same.

But at the end of December the discomfort and the lump seemed to have grown and in the second week of January she made a return visit to her doc who sent her immediately to a cancer specialist. Within a few hours she had another mammogram and then a needle biopsy. Three days later Susan was definitively diagnosed with metastatic breast cancer.

Leaving was clearly off the table.

22. choices

Don't ever move a comfortable cat.

It is very, very wrong. Pomonella was settled squarely on my chest, purring, and it runs against my nature to move a settled and contented cat, particularly an old friend like Po. On top of that it was raining gently on the cabin roof, and there was nothing whatsoever pressing me to get up. Furthermore I was thinking about thinking, and there are few situations more conducive to thoughtful contemplation than the above mentioned circumstance.

On the other hand, it was well past my habitual rising time.

The puzzle jigging in my head was locating the moment of decision. When *do* I decide? The specific decision in question was "Getting Up," one I have mulled before, but the broader question had just suggested itself. *When* do you decide?

When I tell you I was thinking about the decision to get up, I want it clearly understood that I do not mean I was thinking about getting up. Thinking about getting up involves all of the things one wants to do or believes one must do, weighed against comfort and in our culture generally mediated by a clock. The moment of decision is something else again.

I had played with this one sporadically for several years, since I bumped into the notion in an essay by

Douglas R. Hofstadter. He's the sort to pose epistemological conundra that linger and tug at one's philosophical shirtsleeves. He observed that it is impossible to pinpoint the moment at which one decides to get up.

"Oh yeah?" (You say. I said.)

Try it.

You will find, I believe, that it is a very slippery idea indeed. Even if you are one who rigidly obeys an alarm, there is a moment of decision in there somewhere. Even if you think you have previously decided that you will unfailingly leap from mattress to floor at precisely 6:10 a.m. when the buzzer goes off, there remains a hidden moment in which you actually elect to rise.

I had mentioned this to a visiting vacationer two days earlier when the subject of rising time arose, which refreshed the personal challenge. That morning seemed perfectly suited to the experiment. I had no set schedule and my undisciplined lying-in was fully justified to my otherwise Protestantly ethicked workaholism by a late running euchre game with the visitor, presumably also lazing in, who was not due to fly out until late afternoon. That morning I felt sure I could zero in on the pivotal point.

"I will," I told myself, "get up when Pomonella does. I will decide to arise at that precise juncture and the moment will hang suspended in the bell jar of my mental laboratory, ready for close inspection."

At that moment I suddenly generalized the question. Since one is awake when one decides to rise, it is in no wise different than any other decision. This is so idiotically obvious that I almost feel stupid writing it. Almost, but not quite. Because here I suddenly saw why Hofstadter posed the riddle.

During most of our decision making we are busy doing life. Should I walk to work or take my lunch, phone a friend or clean the gutters? That kind of thing. Decision flows from decision and we are too wrapped up in eager anticipation or dealing with consequences to focus very hard on the cusp. But lying abed is different. The old body is idling along. Dreams are fading. Only one decision stands between the horizontal and the vertical, and if any decision might ever be crisply available for study, that one is it.

But when you drill back into your thinking the point of decision endlessly recedes. There is no node.

This realization led to that morning's theory. "All moments of decision are invisible." I feel certain this merely reflects Gödel's proof that no equation can completely describe the system in which it resides. Or, stated another way, you have to be outside of the fish bowl to even know it's a fish bowl.

I was so excited by this idea that I nudged Pomonella off my chest ("Sorry Po."), jumped up, threw on some clothes, popped open my laptop, and began to write. I didn't even start water for coffee until the first four paragraphs of this essay (now reworked into this chapter) were out, the moment of my decision to rise as invisible as ever.

That all happened fifteen or more years ago and I am no closer to achieving consciousness of the cusp of choice that governs rising, or any other choice for that matter. We believe we are rational beings but have very little proof to offer in its defense, apart from some low resistance to magical thinking that a vanishingly small subset of our number call up from time to time.

During the latter half of my partnership with Susan and before her cancer drove her to desperate exploration of every variety of faith healing, we referred to the fantasies amorphously embraced by the label "New Age," as "woo-woo."

In conversation this emerged as "She's into that woo-woo stuff," or, "Sounds pretty woo-woo to me!"

The "woo-woo" wasn't really meant to be harsh or unkind. It was more in the way of gentle sarcasm, triggered not so much by the particular beliefs espoused (since we are all entitled to believe what we will), as by the mercantile slant of many of the practitioners. Sometimes, popular New Age cosmology at the turn of the century seemed like the first fundamentally mail-order religion. Snake oil used to be peddled off the back of wagons, but the business had diversified and gone viral.

The underlying skepticism, however, was more consequential. We are entitled to beliefs, but that doesn't guarantee their truth or utility.

Those who question the dominant paradigm of corporate greed, mercenary wars, boundless consumerism, upward mobility and other pillars of unbridled capitalism seem to split into two camps. On one side of the river, reside the woo-woos. Over here on my side, we practiced the bunny-hug.

Bunny- (or tree-) hugging is the manifestation of a core belief entirely opposite that embraced by the woos. However many attempts are made to bridge the divide, peaceful coexistence involves a large measure of good-natured tolerance. Those who pretend to embrace both perspectives are lost in the fog of a comfortable delusion.

This schism invokes the same issues which spurred Martin Luther to nail his ninety-five theses on the door of his local indulgence-monger. Are we saved by our faith, or by our works? (The further split that triggered Luther's immediate ire concerned whether "works" could include monetary transactions, not dissimilar from the U.S. Supreme Court's confusion of money with speech which makes faithful populists spit nails.)

Orthodox Woos clearly come down on the side of faith. I know generalizations ignore subtle wrinkles, but the bedrock remains: Woos place the inner world first and believe that changing the self will change the rest.

Devout Huggers believe in salvation through work. For us changing the world is physical and political, and the changes in self necessary to achieve that work are also physical and political. Sacralizing work may be useful as a motive force, but in any case, the outer work must be done.

No doubt many Woos are vegetarian bicycling recyclers, while many huggers entertain deep spiritual beliefs, but the practical behavioral divide is as real and deep as a river.

This difference emerged in a conversation with a Woo of my acquaintance. We were discussing the concept of embracing abundance—the belief that the universe will provide for all of our needs if we simply open ourselves to that truth. A simple example of this would be the use of visualization to manifest a loaf of bread.

Then my friend suggested, "Existence is not a zero-sum game."

The idea here is that everyone can enjoy abundance without anyone else giving up anything. Reality is a bottomless cornucopia. We'll make more! I skidded to a halt.

"Wrong," I thought. "It is."

Here is the hurdle: If the world is not a zero-sum game, then faith alone might set us free. If it is, faith will not suffice. In a physically limited system we need to curb our appetites and impose pollution controls. Protecting whole watersheds and building bicycles instead of cars become imperative. In short, we need to work, not meditate, if reality is circular—that is, if the loops of hydrology, nutrients, and energy are closed.

To the Hugger, the Woo embraces pretty illusions which might bring personal joy, but permit the world to die. To the Woo, the Hugger focuses on negative images that block personal joy—and, thus, prevent a perfect world from manifesting.

Environmentalism is the philosophic stance taken by those who believe that we are likely doomed but might be saved by our work; therefore the work must be done. We have no choice.

In everyday life, Huggers and Woos can get along, and do. Both might equally appreciate a sunny fall day, the last flowers blazing in October light, a fresh breeze off the ocean and the spark in loving eyes. They may well agree with Alice that, at the end of the game, we can all be Kings and Queens together. But still, the divide remains. Work or faith?

In reading Thom Hartmann's well-considered and deeply disturbing volume about our oil-dependency, *The Last Hours of Ancient Sunlight* (Harmony Books, 1998), I was brought up short by his conclusion that the most important step in addressing our pending energy-starved doom is meditation and cultivation of "presence" (and the presumed changes in the world that would flow therefrom.) He states, "It's amazing to think that it's possible to change the world by changing ourselves, by changing the way we think and live and experience every moment, but that's been the core message of virtually every religion in history, from the most ancient and primal to the most modern and recent. You can change and save the world by changing yourself."

Well Thom, it may be amazing to think that, but it would be more amazing if anything came of it. Religion has always failed us as a practical approach to problem solving. Magical thinking is magical thinking, no matter how it's done up.

Hartmann even credits research done by the Transcendental Meditation movement which claims proof that crime rates drop in cities when a certain percentage of the population practices TM®. (There wouldn't be any financial motivation behind those ®eported TM ®esults, would there?)

This harkens back to the thoroughly debunked "proof" that prayer by strangers for patients who don't know they're being prayed for affects medical outcomes.

Meditation is a fine practice for those who find it rewarding, as is prayer, but demonstrable success in changing the world is sadly missing. (There is some evidence that we can change our own health outcomes via meditation or laughter, and that those who know others care about their welfare are more likely to recover—but psychosomatic effects are not action-at-a-distance.) Woos give their mystical practices credit when things work out and then allow that the desired outcome must not be God's will when the belief-train leaves the tracks.

As Timothy Ferris noted in writing about the scientific search for other earth-like planets (*National Geographic*, December, 2009), "For thousands of years we humans knew so little about the universe that we were apt to celebrate our imagination and denigrate reality. Now, with advances in science, it has become gallingly evident that nature's creativity outstrips our own."

A deeper difference of opinion embodied in the Woo-Hugger debate involves death and survival. Woos see spirit as separate from flesh and generally believe in some sort of transcendence of this earthly plane— whether that means personal salvation, reincarnation or dissolution into Krishna consciousness and liberation. Such beliefs seem to place a high degree of centrality in *homo sapiens sapiens,* and consequently feed the idea that we are apart from nature, that we are somehow special. At the same time, the idea that the true self will survive death must devalue life. If you believe in a glorious heaven, boundless enlightenment, permanent liberation, why hang around this vale of tears?

Clear-eyed Huggers see our consciousness as a function of our brains and therefore terminal. This life is the only life we will experience, so we'd best make the most of it. Making this world better for everyone, helping those we love and those in need, sharing our joy and ideas and creations, listening to the stories of others, all of it will end far too soon. Time's a wasting!

Furthermore, Huggers' acceptance of science and most particularly evolution cuts hubris down to size. Our species is new in the history of our planet, and temporary. The sun is a middle-aged star. As cosmologist Martin Rees, of Cambridge University, once observed, "Most educated people are aware that we are the outcome of nearly 4 billion years of Darwinian selection, but many tend to think that humans are somehow the culmination. ... It will not be humans who watch the sun's demise, six billion years from now. Any creatures that then exist will be as different from us as we are from bacteria or amoebae."

As will their consciousness. From their distant perspective we will be just one among many species that came and went from this planetary stage, if they are aware of us at all.

23. cancer

Susan was running scared.

And no surprise, as doctors pressed her to opt for immediate radical mastectomy with chemotherapy and radiation to follow. At her surgeon's office I looked at the two mammograms from June and January and could see, plain as day, that the mass now diagnosed as cancer was present in the first. Her surgeon would offer no opinion, demurring with the excuse that he was not a radiologist, but there was simply no doubt that the specialist who gave her an "all clear" in June had overlooked the obvious through incompetence or inattention.

While we dealt with the shock and fear of her diagnosis, the stress drove us further apart. I searched the Internet for information that Susan didn't want to read, found links to experimental treatment methods that were showing promise at various cancer centers and continually tried to be upbeat. The situation approached the surreal. I found myself attempting to lend help and support to my long-term friend and companion in her time of greatest need—at the back end of my decision to exit that abusive relationship which had driven me nearly to contemplate murder.

Susan followed alternative therapy advice and started ingesting a wide range of vitamins and minerals but pursued her allopathic doctors' game plan as well. Over the next two years she would revisit her childhood

Christianity along with reiki, feathering, Rife machines, and artistic therapy, and perhaps others of which I was not aware.

If Susan had survived, I suppose they could have shared the credit with her surgeon, oncologist and radiologist. Therein lies the core problem with verification of alternative cancer treatment regimens—or the mainstream approaches, for that matter. Desperately ill people are often willing to try anything, and do. Some treatments work. Some cancers go into spontaneous remission with no treatment whatsoever. Sorting out the efficacious from the useless is virtually impossible when, in the end, every victim has adopted a unique treatment plan.

By the time she was scheduled for surgery in mid February, we were barely speaking, seeming to differ on almost any matter that came up. She ordered me to keep away from the hospital and enlisted her older brother to come down from Ohio to accompany her during admission and her hospital stay. I complied at the outset, but was in her room waiting with her brother and a friend when she came out of recovery. She was too doped up to complain.

Susan's husband came down from Ohio as well, to help care for her through the first weeks of recovery and she sent me packing to Florida where my mother needed help preparing for a move. Riven by complicated emotions, I wasn't unwilling to follow her lead, and Mom did need my assistance.

A couple of weeks further on Susan started chemotherapy. She permitted me to take her to some of the sessions but reversed course at times and forbade my participation. Her hair fell out. She attended some sort of women's group meetings where she was told that I had caused her cancer. She screamed at me, telling me that I was killing her. She was sick as hell, often vomiting when she managed to eat, and smoking a lot of marijuana to quell the nausea.

Because of her shakes and chills she needed to use electric heating pads, so I disconnected the solar panels and had the house hooked up to grid power. Our solar system was too small to handle extended use of heating appliances.

In August I bought her a new van, feeling that Susan deserved anything she really desired. We had it carpeted throughout, added cruise control, a bed, a stereo system, anything that fit her image of a perfect home on wheels. The summer passed and then the doctors prescribed radiation. Her anger had subsided a little and she let me drive her to some of her radiology appointments. Her lately re-grown hair fell out again. She was really, really sick. She recovered.

During that time she started distributing her stuff to family members—the family treasures and photos she always had intended to leave to them some day. When I questioned it, pointing out that her prognosis was good, and that she had a lot of life left no matter how bad things were, she told me she did not trust me to get things to her family if she died. It came as a blow, given how close I'd become to several members of her clan over the years, not to mention the trust I had vested in her, despite the pain.

In January Susan announced that she wanted to fulfill her lifelong dream of attending Mardi Gras and I made reservations at a great-sounding hotel. We put her canoe atop the van and headed for the Big Easy in time for the holiday week. We bickered and yelled and drank too much and generally made a miserable time of our visit there between parades, but did better during our return trip when we spent some time canoeing in black-water bayous.

During the remainder of the winter Susan argued strenuously for selling our North Carolina home and New

Hampshire cabin and moving to coastal Maine, while I maintained that I had developed a new career in writing and editing and was determined to pursue it in North Carolina.

She insisted that she deserved top consideration due to her cancer bout and demanded that I transfer half of a recently received bequest from my grandfather into her name, which I did. She said she would buy a home in coastal Maine and that she would sell the New Hampshire property to repay me, then we could divide our time between the two and make some decision about residence in the future. Or maybe she'd just move there herself, she wasn't sure. She lit out for New England about the first of May and returned a few weeks later without having found a place on the coast.

In early June she announced that she was headed for Ohio to see her mother who had experienced a spell of poor health. "I'll be back in a week," were her parting words.

The week passed, and then another, and another. Each day when I returned home from work I anticipated seeing her van parked at the top of the hill by the studio. Some time in that third week, seeing the parking spot empty, I suddenly realized that I felt greatly relieved she hadn't arrived. Once I admitted it I realized how tense I had been, fearing her return and not wanting it. I was washed with relief.

She was in remission, she was doing whatever she was doing, and soon I'd be free to follow up on my delayed intention to leave the relationship.

My credit card bill arrived and I saw new charges in Ohio, Massachusetts and New Hampshire, so I realized she'd returned to New England after her visit with her mother. I recall one brief phone call during that summer, and she was indefinite about her plans.

As I relaxed into my time alone I reclaimed my life. For the first time in twenty or more years, it seemed possible to choose for myself without consideration of what Susan might think—everything from when to get up in the morning to how I might spend the rest of my day or my life. I brought my guitar back down to the house from the studio storage loft to which Susan had exiled the instrument and played it on the deck some evenings. After work at the college I stayed late to join a pick-up basketball game or swim laps in the pool. I even signed up for a September trek in Nepal, figuring that either Susan would be home or I would board the cats. The feeling of acting as an independent adult was exhilarating.

Most important of all, I began to cultivate friend-ships. A significant aspect of our lives together had been social isolation. Constant travel, working together and living in the back woods had greatly limited our circle. (I can count on one hand the people from our years together who remain in even occasional touch with me today.) And as I opened up to others and read books about relation-ships, my understanding of the abuse I had lived with became starkly apparent. I resolved that whatever lay ahead I would never acquiesce to abuse again.

Then came 9/11 and another phone call, late in the day. "Did you see what happened?" was her query.

"Yes. Online. With dial-up I've not seen the video. Almost unreal. I don't even know how to react. You still in New Hampshire?"

"Yes."

"Are you headed back soon?"

"No." And she hung up.

Later in the month I was working in Asheville doing charitable carpentry work. The Nepal venture had been cancelled in the aftermath of the terrorist attack. To cut down on commutes I spent alternate nights with friends in town, then back to the mountain to feed the cats and keep up with chores and mail. Saturday, September 22, I returned home in mid-morning to grab necessary tools and to check in with the cats—for some unremembered reason I had spent two consecutive nights in town. That morning we'd pulled a toilet in order to fix a rotten floor and as it was the only toilet in an occupied residence, I needed to get back pretty quickly to finish the carpentry so it could be reset.

There was the van.

I parked, took a deep breath, and headed down to the house.

"Welcome home!" I called as I crossed the deck.

No reply. And no one downstairs. I went to the stairwell, "Welcome home!"

"Where the fuck were you?

"Working in town."

"Why didn't you have dinner ready when I got home Thursday?"

"Are you serious? Why didn't you tell me you were coming home?"

"Asshole!" she spat back.

Seeing that the shouted conversation was going nowhere, I climbed the stairs and said, "I'm glad you're home, but I've got to get back to town, there's a toilet that has to be reset. But I'll pick up whatever you want and head right back and fix dinner."

"Trout," she replied. "Asshole."

"Okay, trout. By the way, you're looking great and I'm looking forward to hearing about your summer."

When I returned home several hours later the van was gone. No note. I worked that evening at my computer, editing the Warren Wilson environmental journal, and e-mailed a friend about Susan's angry reception.

My friend's response was, "I guess you didn't get a

blow job, either," which may be crude, but was, as intended, pretty funny.

The next morning I returned to the job and then home again in late afternoon. Susan was back and had read the e-mail. Her insistence on reading all of my incoming and outgoing e-mail and snail-mail over the years is one of many ways she managed my life. She was screaming accusations when she came out of the studio— alleging that I'd been having affairs while she was gone, that I was a scum for treating her that way when she was ill, and so forth.

I managed to maintain my cool and told her, "Think what you choose. But if you don't like what you read in my e-mails, I'd say you should quit reading them."

She was spitting in my face as she continued her screams and I shrugged. "Believe what you like Susan. Let me know when you're interested in dinner. I brought home trout yesterday. Where'd you go?"

"None of your business."

Susan went down to the house and I went into the studio, where the sinful e-mail was up on the screen of my iMac. I worked for a couple of hours and then headed down the hill to fix dinner.

When she heard me come in, Susan came downstairs, calm but icy. "Something has changed about you." Her cadence was deliberate and her tone accusatory. "What is it?"

Lowering myself into a chair I replied, "Susan, I've had a lot of time to think about us and about myself this summer, and I've decided I don't need you to run my life anymore. I'm a grown-up, I can manage myself."

"I have never tried to run your life," she answered.

"Fine. Then we're agreed. We're good to go."

"I never ran your life!" she said with more force.

"Okay. I was wrong. You never ran my life, so you

won't mind never running it again. That's okay with me."

She cursed me a few times and returned upstairs while I busied myself fixing dinner. When I'd served the food I called her down and we ate in silence for several minutes. Then she picked up a nine-inch kitchen knife I'd placed beside a chunk of cheese on a cutting board, examined it while she held it vertically and fixed me with a glare.

"You have to sleep some time," she warned.

I laughed and shook my head. "You must have had a great time this summer. You sure came back with a new attitude."

Little did I suspect, as I met her angry glower, that it would be my last and lasting memory of her face. She returned upstairs after dinner and I never saw her again.

After cleaning up the kitchen I returned to the studio and tried to work for a few hours. But her threat kept turning round in my head. Given her screamed accusations coupled with the entirely unreasonable expectation that I could have somehow anticipated her unannounced return, it wasn't clear to me that I was dealing with a rational individual. I thought back to my own homicidal temptation two years earlier and understood very clearly that our thoroughly dysfunctional relationship veered toward the crazy-making. Part of me wanted to be brave, to go to my bedroom and show her both that I wasn't scared and, in a way, that I still trusted her. Another part of me said that was an incredibly stupid idea and that I shouldn't go anywhere near the house. That part won.

I briefly considered sleeping in her van, but it was locked, so I decided to return to Asheville after phoning to be sure I'd have a place to stay. Considering Susan's anger about the e-mail and her long-standing disparage-ment of my writing work, I unplugged my computer and carried it to the truck. No use taking chances. Then I tiptoed down to the house to recover my guitar.

I returned home Monday night to a dark house, and I was locked out. I knocked but if she was there she didn't choose to answer. I didn't have a key, since we had never locked it unless we'd be gone for an extended time. I left a note on her van which was also locked. On Tuesday evening I returned to find her gone with the house once again locked tight. I went into the studio and tried to use the phone, but it was dead. Investigating, I saw that she had ripped out all the wires from the junction box under the studio. I was glad I'd removed the computer.

I got into the house by climbing an extension ladder to an unlocked upstairs window and found she had taken all the keys to the house from the key dish atop the fridge. The cats were missing as well.

I sized things up and weighed the advice I'd received from friends based on their experience of divorce and property fights. I loaded up a couple of boxes of letters, heirlooms and photographs, my writing files, my favorite books, the art work I considered mine, a suitcase full of clothes and then filled the truck bed with my tools. I rushed through the effort, as if I were fleeing an oncoming forest fire or flood, anxious lest I forget something I'd later regret.

My friends very kindly made space for me and were uncomplaining about my need for a landing place while I tried to sort things out. I was loathe to rent or lease, since I figured I'd soon be back in the house. It was, after all, my home, and Susan had never considered it hers. So, if we were now going to split, it just made sense that it would be part of my share. But I definitely didn't want to sleep there in the interim, her threat and her grim visage working in my mind.

Finally I recognized that any settlement would be a while in coming and rented an apartment in Asheville and transferred my phone number from the house to the apartment. Not long afterward Susan phoned with news of a medical check-up. She was crying. "It's metastasized to my brain," she half-whispered.

Knowing that diagnosis was almost certainly a death sentence, I started crying myself, "Oh, Susan. What can I do for you? I'll take care of you at the house if you want. Or, if you don't want me near I'll get nurses. Whatever you need, just let me know."

Click.

And that was the last time I spoke to her.

The next day my phone was cut off, I'd forgotten that her name was on the decades old account, though I'd always paid the bills (and had been the one who even wanted a phone in the first place). She had demanded that the phone company deny me the number and so cancelled the account. At about the same time a friend in Ohio let me know that Susan had returned there and had rewritten her Will.

That meant trouble. For all our unmarried years together our Wills had left everything other than family keepsakes to each other. Something was up. A week or so later I received a letter from an attorney ordering me not to return to Susan's house, and indicating that she wanted to reach a reasonable settlement with me in division of our joint property. I hired an attorney to respond and her attorney assured mine that Susan intended to settle agreeably.

In January a friend saw a legal notice indicating that my Black Mountain home had changed hands. Sure enough, she had given it to a nephew. Alarmed, I called the county clerk in New Hampshire to learn that our property and cabin there had been sold as well.

The memories thus triggered were bitter:

When we purchased property in New Hampshire in 1979 she said, "The property should be in my name, that way if there's ever a law suit concerning your construction work our place will be safe. You can trust me, I will always be fair."

When we purchased property in North Carolina in 1982 she said,:"The property should be in my name. You can trust me, I will always be fair."

When I purchased a new Dodge van in 2000 she said, "The van should be in my name. You can trust me, I will always be fair."

In 2001 she said, "Transfer half of your grandfather's bequest into my name. I will sell the Deerfield property and pay you back. You can trust me, I will always be fair."

Then, in April, 2002, came the offer of a "fair settlement" via Susan's attorney, Howard Gum. It amounted to this: "You can have 10 percent of everything we saved jointly over twenty-five years, I will keep 90 percent, and, oh, I'll keep half of your grandfather's bequest, too." Adding insult to injury, I'd later learn that Gum's law partner handled the transfer of my house to her nephew at the same time that Gum assured of Susan's equitable intentions.

At the same time, it felt terribly tragic. I could see that her extraordinary paranoia and anger could easily have been exacerbated by the brain cancer, and now she was dying. I learned from a friend that she had moved in with relatives in Ohio. Her family, having been filled with her allegations of infidelity and who knows what other imagined crimes, cut me off completely—other than a single hateful letter from her brother in Oregon.

When Susan died in late May they didn't even bother to let me know she was gone.

24. perspective

We are wanderers, you and I.

We're on a too-short sojourn from wasn't to isn't, on a trajectory we were thrown into, on a pathway we invent. Our footsteps are tiny and even the most dedicated walkers among us do not travel far between birth and death. We tend to focus on the local landscape, on the immediate event, and to whatever extent we see beyond depends in large part on the stories of others. A few have gone up the mountain to peer into the next valley, a few have crossed the ocean and returned, and now a very few have looked back from space.

The world seems vast to a walker with a two-foot, two-footed stride. The sky seems high and the oceans deep to a two-meter being looking upward and down. Only the insight of ten thousand generations' tales can begin to instruct us in the truth of it all. Our biosphere is a thin skin on a wee world near a middling star in an average galaxy in a normal cluster, among billions. It is only by the greatest good fortune that we have come to be. It is only in this one known place in the cosmos, in this one brief interval in the long stretch of time, that stellar radiation and chemical recombination have erupted into life. It is only in this thin skin of air and water, no thicker than the varnish on a classroom globe, that life emerged, photosynthesized, ate, mated, dreamed and spoke. "We are wanderers, you and I ..."

It is only in this last flickering moment that we wanderers have set ourselves to meddling with the biology and chemistry and nuclear forces that have balanced their own books to create our living planet. It is only in the last quarter of a moment that there have been enough of us for our meddling to have much effect. And it is only in the last blink of our eye that we have begun to understand that our numbers and our meddling can effect systemic change in the fragile bubble of life that is our world.

We have built our empires on petroleum, the stored energy of a hundred billion sunrises, and are using up that energy in less than one hundred thousand days. We have permutated organic chemicals to do our bidding as pesticides, solvents and fuels, and have introduced those twisted mutagens into every living cell on earth. The day I wrote these words I heard news that a patent has been granted for the five millionth chemical compound.

Five million new chemical compounds introduced to our world, and not a clue how most of them interact, catalyze each other, or disrupt living cells.

We have punched holes in the sky and fermented dead zones in the oceans. The pharmaceuticals in our urine build up in estuarine fish and tests of whales around the globe reveal heavy metals and chemicals at levels that qualify them as toxic waste sites. We have soured the rain and ripped away the topsoil where the magic of life must set root.

Perhaps we see more clearly now, and perhaps there is time for change. Not every kid with a chemistry set blows up the basement, after all. But the changes must be rapid and widespread if we are to undo our deviltry before it undoes us. The infusion of endocrine disrupter and hormone mimetic chemicals which first came to broad awareness in the past two decades is endemic and getting worse. New studies reveal a frightening correlation between pre-natal exposure and low intelligence, reduced attention span, aggression and reproductive dysfunction. To the reports of mutant frogs, homosexual seabirds, immune deficient cetaceans, and sterile alligators we now add declining sperm count and increased testicular cancer in humans. Our chemical assault is undoing us. Basement bombs are small potatoes compared to the decoupling of our genetic freight train.

There exists the possibility that the human genome project will finish building its Secret Decoder Ring just in time to discover that we have already monkey-wrenched the works.

Perhaps there is time for change. We are already changing. The urge for organic food reflects a populist hunger for a step back from the brink. The surge in SUV sales answers with a lemming-lunge for the cliff. Developing nations at long last embrace population reduction. The U.S. government funds development of plants that don't produce viable seeds while conservative Christian elites decide that big families are back in fashion. School children want to save the whales, Makah warriors and Japanese gourmands want to spear them. One step forward. Two steps back.

But, forward and back from what?

If you pull your cell phone out of your ear for five minutes and focus your gaze on a fast moving stream you would see standing waves. Waves were standing there before you bothered to notice, and they will be standing there long after you exit this plane, but at the moment they might just be able to tell you something. Something important.

First off, they don't exist.

Standing waves are a pattern without substance. They emerge as a result of water flowing over the river bottom and being channeled in coherent patterns. A form emerges. A wave.

But the water keeps right on flowing, and the thing we see is not a thing at all. It is the idea of a thing, the energy of a thing, but is composed of a quickly changing stream of water molecules which join in the local dance for a moment and move along. If you watch for a while, you will see that some continue alone, while others merge and redivide once more with the pulse of the river, what Paul Simon called "the rhythm of the saints."

Scientists are exploring this phenomena in the study of chaos these days, and the ebb and flow of noise and music seems to explain a lot about everything. (Noise being incoherent sound, and music its opposite.) Noise tends to organize itself into music and devolve into noise again. Dust gathers to become a star and burns itself back into dust, or gathers itself to walk on two legs in two-foot steps before devolving into dust and scattering to the wind once more.

The molecules which constitute your body are only stopping by on their way to somewhere else. No doubt you've heard, or figured out for yourself, that your body renews itself over days and months and years, building fresh cells out of the stuff you eat and breathe, and dumping the used matter to be recycled elsewhere. Perhaps you haven't thought of yourself as a standing wave, but you are: albeit in flesh that appears to be a bit more solid and stable than a whitewater rapid. In that sense, you too do not exist. You are simply a pattern.

As Eiseley noted concerning fossils on his wall and desk, "Just once out of all time there was a pattern that we call *Bison regius*, a fish called *Diplomystus humlis*, and. at this present moment, a primate who knows, or thinks he knows, the entire score."

Moreover, quantum theory has proven that energy and matter are interchangeable, that an observer changes what is observed in the act of observing, and most astonishing of all, that when a particle of light is split in two and the spin of one particle is reversed, the twin reverses instantaneously, at any distance.

But this concluding chapter is not about standing waves or quantum mechanics or the riddle of existence, it is about violence in schools, militia groups, computer networks, radio signals and the end of life as we know it. I suppose that means it also about standing waves and the riddle of existence, but I didn't want to confuse you.

One thing you need to remember about standing waves is that if you were to wade out into the stream with a wrecking bar and dislodge a few boulders, the wave would change. It would seem to disappear, reemerge and then stabilize in a new form. Even the shifting of smaller stones will change the shape of the wave, and pulses reflected from other parts of a rapid will blip up as random splashes and swirls within the more stable framework of the form.

Similarly if a few of the boulders in your life had been different, you would display another pattern than you do right now. If you had or hadn't been abused, exposed to ionizing radiation, dropped on your head or been chosen last, or first, for basketball, you would be different than you are today.

I am who I am, for better or worse, thanks to genetics and parenting, the inter-parental role modeling I experienced, my sibling and cohort relationships, the privileges accorded my race in a wealthy nation and the educational opportunities I have taken up or eschewed. The early, primal experiences profoundly affect all that follows, of course, and altogether they have formed the belief system through which I gauge the world.

All that I am drew me into a long-enduring relationship which I defined in retrospect as abusive— with myself as the abused party. It's quite natural to see ourselves as the central players in the universe and to experience our own pain as the most significant. After all, we feel it. We are the ones being hurt. We are the ones who do and have done to us.

The story of the other is harder to grasp, particularly when it is close at hand. Empathy in the abstract comes easily to those of us who aren't innate sociopaths or narcissists, but empathy for the person who is breaking one's heart comes hard.

In the years since Susan died my understanding of our story has changed. First came the admission to myself that I had suffered abuse. When recognition of pain had popped up from time to time over the years it had been quickly quashed with a comforting blanket of denial.

Then came a period of time in which I allowed forgiveness to temper the anger. I embraced the idea that she had played out her life as best she could, given her own primal experiences and make-up. I could then believe that she hadn't intentionally hurt me.

With time and introspection I have come to a place where I understand that I hurt Susan, too. She was deeply hungry for love, for affirmation, perhaps for the Daddy she had lost, perhaps for something further back, some feeling of abandonment in a Dr. Spock inspired bassinet, a bottle instead of a breast, a childhood spanking or some deeply repressed sexual abuse. I wasn't emotionally available, and if her response was distancing, mine didn't act to close the distance. Both of us were trapped in righteousness.

All of the above is true for human society as well.

The stuff of life moves along like a river and the individual need for food and shelter and clothing and sex and meaning combine to become the group needs of a family or a tribe or a nation. Depending on the underlying riverbed, the combination rises up to form standing waves: the Roman Empire was a biggie, a pattern that endured for nearly a millennium. China presents an even longer stability, a slowly shifting swell that has moved around the riverbed over its 40-something centuries. Christianity is another long-term standing wave—shorter by far than Hinduism, Buddhism or Judaism, but five hundred years ahead of Islam and fifteen hundred ahead of the Mormon wavelet. Liberal capitalism is a youngster, as is the United States.

From time to time the boulders underlying our social river are moved by natural or human intervention.

Epidemics have repeatedly dislodged the standing waves of social organization. When the Black Death killed off roughly half of Europe's population early in the past millennium, feudalism and Catholicism took it on the chin. In the ensuing labor shortage, the first strikes occurred. ("Take this job and shove it. Bring on the Wobblies!") And, as "Geologian" Thomas Berry once observed, one response to the unfathomable suffering and morbidity was to believe that God had decided to punish sinners, which seemed to include devout Catholics. Martin Luther nailed some questions on a church door and "Bingo!" the Protestant Reformation splintered Christianity into its many modern veins.

Recent research suggests that medieval plagues even caused the Little Ice Age that chilled our planet from 1350 to about 1600. The death of so many people permitted cropland to reforest which pulled enough carbon out of the atmosphere that the climate cooled. Carbon levels rose as Europeans repopulated North America—where indigenous peoples had been wiped out by plague and small pox—and cleared vast tracts of woodland. Nice weather returned.

Technology has often done the same mischief to our social order: as when the longbow ended the military advantage of a mounted cavalry, or knitting machines replaced hand looms.

But back to the rocks in our present river. We have moved a whole lot of social boulders in the past century and long stable patterns are everywhere in flux.

We introduced universal mobility, bringing the historically gradual migration and intermingling of peoples to a crescendo—almost anyone can go almost anywhere. We decoupled sex from procreation, and decoupled procreation from marriage. We conquered many diseases and provided a cornucopia of food without reducing the birth rate and more than tripled our numbers. We have embraced philosophies which argue for the ultimate meaninglessness of our lives, while simultaneously embracing celebrity as the pinnacle of success. We made world wars possible, then unthinkable, and while cobbling together a framework for world peace turned weaponry into a cornerstone of international trade. We instantized information and turned relevance on its head: regional conflicts on the other side of the globe can seem more important than a local event which materially affects one's life—distorted by TV's electronic lens. We have turned a quiet world soaked in the sibilance of wind, waves, crickets and birdsong into a place of never ending mechanical racket: jets and autos and trucks and trains, refrigerator hum, air conditioner whoosh, clock tick, phone bleat, hard drive whir and the pervasive urban thrum of car stereos on steroids.

In the chaotic aftermath of such boulder shuffling, everything seems to be in flux. What appears to be happening is the emergence of a new standing wave called Globalism, but what it might ultimately look like is anybody's guess. We might be leaving behind the rootedness and unwavering familial bonds which have changed very little since we became the genus *Homo*. Or, we might be breaking down the larger social groupings of nation and faith into scads of separate cells, more locally rooted and bound than ever before.

Just as the United Nations seemed poised to become the Mother Church of world peace, ethnic feuds and terrorism exploded on every corner, and reactionaries, fearful of the uncertain New World Order, spread rumors of black helicopters, secret road signs and foreign invasion.

Just as universal literacy seems at least possible, agreement about what the literate should be permitted to read has fractured, and the very necessity of reading is questioned. Science books are rewritten to include religious myth as putative fact. Home schoolers design their own curricula and Moslems and Christians burn books that offend their often similar sensibilities. China bans import of some titles while exporting bootleg versions of a thousand others and censors its internet services while becoming the world's principle supplier of the computer parts that make the internet possible. A Pax Americana tentatively emerged after the standing wave of World Communism melded into the capitalist fold, and the American response has been to become the world's biggest arms dealer.

I think we should not be surprised that children are shooting each other, building bombs and killing themselves at record rates. Saddened, yes, but not surprised. Just as the macro-terrorism expressed on a Pan Am flight over Lockerbie, at a Federal building in Oklahoma City, in a Tokyo subway or in the tumbling Twin Towers reflects our unsettled and unsettling era, the micro-terror at Columbine High School, Virginia Tech and other schools is reflective of the confusion. As the standing waves of nation and ideology have collapsed, transnational corporations have rushed into the void, and money-changers are making or breaking national economies with 24-7 arbitrage sweeping the world market. Bubbles rise and bubbles burst.

Faith, too, is strained, with E.T. cults, Branch Davidians and fundamentalists choosing mass suicide as a consummate declaration of belief while suicide bombers practice their religious rites on a one-off basis.

Megachurches rise and megaministers fall. The Family insinuates itself into the highest realms of government and wiccan covens spring up like mushrooms in the fertile countryside.

Humans exhibit a longing for stability, which, if history is a guide, will soon congeal in another durable equilibrium. The longing may be an organizing principal begat by chaos, or simply a practical means of assuring regular meals. There are weapons aplenty and plagues are on the rise. How long will we know chaos before the next standing wave arrives?

Here's the puzzle. We don't exist but we do. What we see may depend on what we believe, but what we see together is better explained by science than faith. Individual belief begets visions, untestable by definition. Repeatable experiments yield predictable interpretations and potential solutions. As Francis Bacon observed at the outset of the scientific revolution, when inquiry begins in certainty it is likely to end in doubt, whereas when we enter in doubt the truth will often emerge. If we choose to coexist we need to embrace agreed upon solutions, not fissiparous fantasy.

Still wandering we are, and unsure of our goal. Once we weren't, and soon we won't be. Here in the middle we tread our narrow path. The waters have not yet begun to cohere. The music we hear is still noise.

Yet deep beneath the waves of the three quarter portion of our planet which is under water, unfallen whales are singing.

Index

Acknowledgments

I am greatly indebted to Alex Cury who has acted as sounding board and advisor for this work and much else over the past several years. Absent her friendship and wisdom my understanding of my past and present, indeed my understanding of my understanding, would be much the poorer.

More directly, Alex contributed her editorial skills to this work and my biographies of Billy Graham (2007) and Bobby Lee Medford (2008), always to good effect.

My longest-enduring friend, Philip R. Breeze, lent his thoughtful and well-practiced editorial hand as well, across 4,500 miles from Oahu, thanks to the magic of the World Wide Web. His early and lifelong fascination with language has ever been an inspiration, despite the petroleum-fueled gaps in our acquaintanceship through the years.

Readers Charlie and Laura Thomas offered editorial advice and psychological insight, contributing to both my work and my life in ways hard to quantify or describe.

Brian Sarzynski was particularly helpful in the realm of linguistics, together with wide ranging discussion of many of the topics assembled here.

Ron Ogle, gatekeeper of many doors of perception, knows more than he admits, and has turned over stones to reveal snakes and gems at critical moments over several years.

Between them all I have managed to avoid various stupidities, typographic collisions, factual errors and logical fallacies that would otherwise make me look much worse than I do. Whatever errors and idiocies remain are mine and mine alone. Thanks, again, to all.

I feel drawn to thank Susan as well, though post-humous thanks seem to me to be of little worth. Despite the pain, there was much pleasure, and if I learned some things late, at least I learned and am learning. Fifty-five seems much too young to exit this blue ball and the eight years since your passing seem no time at all. If, against all odds, there actually *is* a river to skate away on, I hope you've taught your feet to fly.

About the author

When Cecil Bothwell was elected to Asheville's City Council in 2009 a Confederate, Christian activist attempted to block his induction to office, based on an archaic provision of the North Carolina Constitution which bars from office anyone who "shall deny the being of Almighty God." The trigger was Bothwell's statement concerning personal belief in his critical biography, *The Prince of War: Billy Graham's Crusade for a Wholly Chrisitian Empire* (Brave Ulysses Books, 2007).

The story went viral, reported and repeated around the globe in eight languages, on television, radio and in print. The debate concerning church/state separation erupted in blogs and letters-to-the-editor pages, and hundreds of supporters flooded the author with snail-mail, e-mails and donations to his campaign fund.

Bothwell is an investigative reporter and biographer based in Asheville, and has received national awards from the Association of Alternative Newsweeklies and the Society of Professional Journalists for investigative reporting, criticism and humorous commentary. Former news editor of *Asheville City Paper*, former managing editor of Asheville's *Mountain Xpress* and founding editor of the Warren Wilson College environmental journal *Heartstone*, he served for several years as a member of the national editorial board of the Association of Alternative Newsweeklies and currently serves on the boards of two international educational nonprofit organizations working in Latin America. His weekly radio and print journal, *Duck Soup: Essays on the Submerging Culture*, remained in syndication for 10 years.

He blogs at: bothwellsblog.wordpress .com